Nuclear Magnetic Resonance Spectroscopy - Recent Research and Applications

Edited by Shagufta Perveen

Published in London, United Kingdom

Nuclear Magnetic Resonance Spectroscopy - Recent Research and Applications
http://dx.doi.org/10.5772/intechopen.1007235
Edited by Shagufta Perveen

Contributors
Aleš Mohorič, Carmen-Irena Mitan, Carmen Jiménez-Espinoza, Deniz Uner, Ecem (Volkan) Deveci, Emerich Bartha, Francisco Marcano Serrano, José González-Mora, Petru Filip, Robert Moriarty, Shagufta Perveen

Notice

Statements and opinions expressed in the chapters are these of the individual contributors and not necessarily those of the editors or publisher. No responsibility is accepted for the accuracy of information contained in the published chapters. The publisher assumes no responsibility for any damage or injury to persons or property arising out of the use of any materials, instructions, methods or ideas contained in the book.

First published in London, United Kingdom, 2025 by IntechOpen
IntechOpen is the global imprint of INTECHOPEN LIMITED, registered in England and Wales, registration number: 11086078, 167-169 Great Portland Street, London, W1W 5PF, United Kingdom

For EU product safety concerns: IN TECH d.o.o., Prolaz Marije Krucifikse Kozulić 3, 51000 Rijeka, Croatia, info@intechopen.com or visit our website at intechopen.com.

British Library Cataloguing-in-Publication Data
A catalogue record for this book is available from the British Library

Nuclear Magnetic Resonance Spectroscopy - Recent Research and Applications
Edited by Shagufta Perveen
p. cm.
Print ISBN 978-1-83635-220-4
Online ISBN 978-1-83635-219-8
eBook (PDF) ISBN 978-1-83635-221-1

If disposing of this product, please recycle the paper responsibly.

IntechOpen

intechopen.com

Built by scientists, for scientists

Meet the editor

Dr. Shagufta Perveen, Ph.D., FRSC, is a Senior Scientist and expert in phytochemistry and analytical chemistry with over two decades of experience in natural product research. She earned her Ph.D. in chemistry and has supervised numerous MSc and Ph.D. students. Her work focuses on structure elucidation using advanced techniques, including LC-MS/MS, NMR, UV-Vis, and IR spectroscopy. Dr. Perveen is currently based at the University of Wisconsin–Madison, where she is working on many projects involving metabolomics, chemometric analysis, and AI-assisted data interpretation. A recipient of several academic honors, she is a Fellow of the Royal Society of Chemistry (UK) and actively contributes to international collaborations in drug discovery and biomolecular research.

Contents

Preface

This volume provides a comprehensive overview of the principles, methodologies, and applications of Nuclear Magnetic Resonance (NMR) spectroscopy, specifically designed for researchers, educators, and advanced students in the fields of chemistry, biochemistry, and materials science. NMR remains one of the most powerful and versatile tools for determining molecular structures, studying dynamics, and elucidating interactions in various chemical and biological systems. With the continued advancement of both high-field and benchtop instrumentation, as well as computational and analytical approaches, NMR spectroscopy is evolving rapidly, becoming more accessible, more precise, and more integrative than ever before.

The volume opens with an introductory chapter providing foundational concepts in NMR spectroscopy, setting the stage for both newcomers and experienced users seeking a concise yet thorough refresher. Subsequent chapters delve into specialized areas of research, including chemical exchange dynamics and the use of variable-frequency CPMG echo trains, operando NMR spectroscopy using benchtop instruments, and innovative approaches to structural analysis, such as the three-sphere method for calculating dihedral and tetrahedral angles in cyclic systems. A dedicated chapter further explores the crucial role of NMR in biomolecular chemistry, underscoring its significance in studying macromolecular structures, interactions, and dynamics.

Each chapter is authored by experts in their respective areas, providing both theoretical context and practical insight. This edited volume reflects the interdisciplinary nature of NMR spectroscopy and its expanding utility across academic and industrial research. By bringing together classical principles and cutting-edge developments, the book serves as a valuable resource for those seeking to deepen their understanding or explore new frontiers in NMR-based analysis.

I sincerely hope this compilation will inspire continued innovation and foster deeper engagement with the fascinating world of NMR spectroscopy.

Shagufta Perveen
University of Wisconsin–Madison,
Madison, Wisconsin, USA

Chapter 1

Introductory Chapter: Nuclear Magnetic Resonance Spectroscopy

Shagufta Perveen

1. Introduction

Nuclear magnetic resonance (NMR) spectroscopy is one of the most powerful and versatile analytical tools used in chemistry, biochemistry, and materials science. Based on the absorption of radiofrequency radiation by nuclei in a magnetic field, NMR provides unparalleled insights into molecular structure, dynamics, and interactions at the atomic level. Over the last few decades, NMR has evolved from simple 1D spectral techniques to multidimensional, high-resolution, and in vivo imaging methods. With the advent of cryogenic probes, hyperpolarization, benchtop systems, and AI-assisted data processing, NMR remains at the forefront of both fundamental research and industrial applications (**Figure 1**).

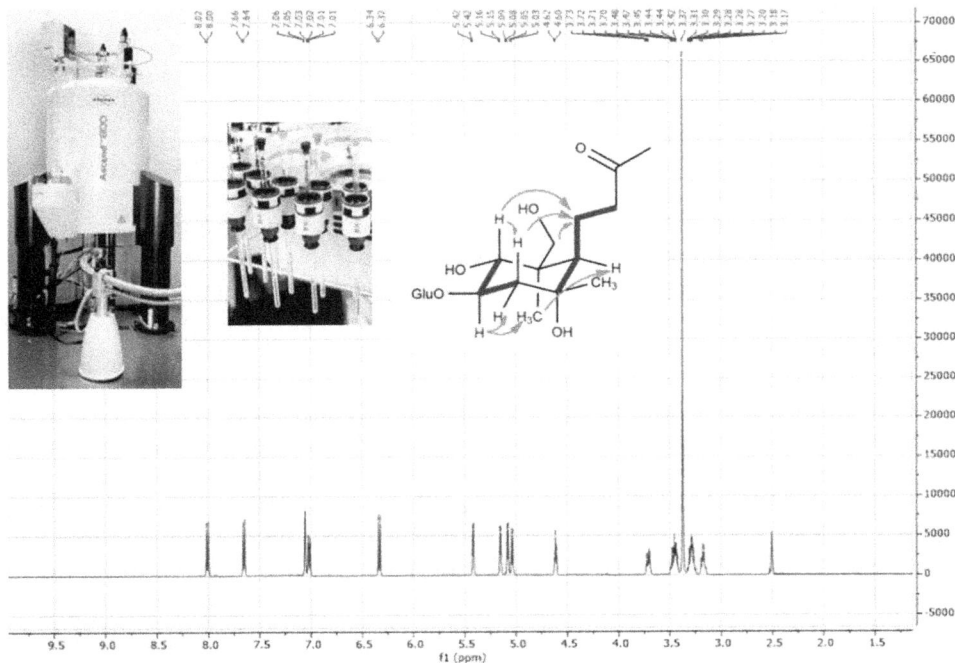

Figure 1.
Basics of nuclear magnetic resonance spectroscopy.

IntechOpen

2. Basic principles of NMR

NMR operates on the principle that certain nuclei (^{1}H, ^{13}C, ^{15}N, ^{19}F, ^{31}P) possess intrinsic spin and magnetic moments. When placed in a strong magnetic field (B_0), these nuclei can align with or against the field. Upon application of radiofrequency (RF) pulses at a resonant frequency (the Larmor frequency), these nuclei are excited to higher energy states. As they relax, they emit signals detected and translated into spectra that reveal molecular information such as chemical environment (chemical shifts), coupling patterns (*J*-coupling), and spatial proximity (NOE).

3. Recent methodological advancements

3.1 Cryogenically cooled probes

Cryoprobes significantly enhance sensitivity and reduce noise in NMR experiments. This has enabled the acquisition of high-resolution data from minute sample quantities, particularly useful in protein NMR and natural product structure elucidation. In cryoprobes, the probe electronics and preamplifiers are combined into a single unit and cooled to cryogenic temperatures. Cooling the probe slows the random motion of electrons inside the electronic components, which reduces the electronic noise, or static, that they introduce into the NMR signal [1].

3.2 Hyperpolarization techniques

Hyperpolarization methods such as dynamic nuclear polarization (DNP) and parahydrogen-induced polarization (PHIP) have drastically increased the sensitivity of NMR by several orders of magnitude. These methods are being applied in real-time reaction monitoring and metabolic imaging. Hyperpolarization techniques in NMR enhance nuclear spin polarization, leading to significantly boosted NMR signal sensitivity. This is achieved by preparing the nuclear spin system in a state with an increased population difference, going beyond the thermal equilibrium achieved at typical NMR measurement conditions. Several methods, including DNP, PHIP, and optical pumping, are employed to achieve this enhanced polarization [2].

3.3 Benchtop NMR

The miniaturization and affordability of benchtop NMR spectrometers have democratized access to NMR technology in teaching, quality control, and field-based research. Benchtop NMR refers to a compact and portable version of NMR spectrometer, designed to fit on a laboratory bench. These spectrometers use permanent magnets, which are smaller and more affordable than the superconducting magnets used in traditional NMR. It offers a cost-effective and convenient alternative for various applications, including research, quality control, and educational settings [3].

3.4 Artificial intelligence and machine learning

NMR is essential for identifying and understanding small molecules, and artificial intelligence (AI) techniques, particularly machine learning, have improved the accuracy and efficiency of data analysis. By integrating AI, increasingly complex datasets can be handled. Artificial intelligence (AI) and machine learning (ML) are revolutionizing NMR research by automating complex tasks, enhancing data analysis, and enabling new insights into molecular structure and dynamics. AI/ML applications in NMR include peak assignment, spectral simulation, data processing, and even predicting molecular properties. AI tools now assist in peak deconvolution, structure prediction, spectral assignment, and even spectral acquisition optimization, saving significant time and increasing accuracy [4].

4. Applications across disciplines

4.1 Natural product chemistry and metabolomics

NMR is crucial for structural elucidation of new natural products and profiling complex metabolomes in plants, microbes, and humans. NMR spectroscopy has become an indispensable tool in metabolomics, particularly in the analysis of natural products. Its nondestructive nature and ability to provide detailed structural information make it ideal for profiling complex biological samples. A recent study highlighted the application of NMR in plant metabolomics, where researchers developed protocols to analyze the vast chemical diversity of plant metabolites. By utilizing ^{1}H NMR spectroscopy, they were able to identify and quantify a wide range of metabolites, facilitating a deeper understanding of plant biochemistry and aiding in the discovery of bioactive compounds. Such advancements underscore NMR's pivotal role in advancing natural product research and metabolomic studies [5].

4.2 Structural biology

NMR spectroscopy continues to be a vital tool in structural biology, offering unique insights into the structure and dynamics of biomolecules. A notable recent advancement is the integration of NMR with AI to expedite protein structure determination. For instance, Bruker has demonstrated how combining ultra-high-field NMR instruments with AI-based software solutions enables the precise identification of binding molecules and characterization of interactions. This approach has revitalized NMR's importance in early drug discovery programs, allowing for the resolution of protein structures in a timeframe competitive with other techniques like X-ray crystallography and cryo-electron microscopy. Multidimensional NMR enables determination of 3D structures of biomolecules in solutions, offering insights into protein folding, ligand binding, and conformational dynamics [6].

4.3 Clinical and diagnostic applications

NMR spectroscopy has become an indispensable tool in clinical diagnostics and personalized medicine, particularly through its application in metabolomics. By enabling the comprehensive profiling of metabolites in biological fluids like blood and urine, NMR facilitates the identification of disease-specific biomarkers and enhances our understanding of metabolic alterations associated with various health conditions. Recent studies have demonstrated NMR's efficacy in detecting unique metabolic signatures linked to diseases such as cancer, cardiovascular disorders, and neurodegenerative conditions. These insights are crucial for early diagnosis, monitoring disease progression, and tailoring individualized treatment strategies. Moreover, the noninvasive nature and high reproducibility of NMR make it particularly suited for longitudinal studies and routine clinical assessments, underscoring its growing role in advancing precision medicine [7].

4.4 Materials science

Solid-state NMR reveals atomic-level structures and dynamics in polymers, catalysts, ceramics, and energy materials. Solid-state NMR spectroscopy has become an indispensable tool in materials science, providing atomic-level insights into the structure, dynamics, and electronic environments of a wide array of solid materials. Recent advancements have significantly expanded its applications across various domains. For instance, in the study of hybrid metal halide perovskites (MHPs), which are pivotal for next-generation optoelectronic devices, solid-state NMR has been instrumental. Researchers utilized multinuclear solid-state NMR to investigate structurally diverse low-dimensional MHPs, revealing critical information about their local structures and dynamics that influence their optoelectronic properties. In the realm of catalysis, solid-state NMR has been employed to characterize active sites in titanium dioxide (TiO_2), a widely used photocatalyst. This technique provided detailed insights into the local environments of titanium centers, enhancing the understanding of their catalytic behaviors [8].

4.5 Environmental and food chemistry

NMR spectroscopy has become a powerful analytical tool in environmental sciences, offering nondestructive, high-resolution insights into the molecular composition of complex environmental samples. It is widely used to study soil organic matter, water pollutants, atmospheric aerosols, and plant metabolites affected by environmental stress. For instance, NMR is crucial for analyzing dissolved organic matter (DOM) in freshwater and marine ecosystems, helping researchers assess carbon cycling, water quality, and ecosystem health. Solid-state NMR techniques also allow for the characterization of humic substances and soil carbon pools, contributing to our understanding of soil fertility and carbon sequestration. A recent application involved ^1H and ^{13}C NMR to monitor microplastic degradation and track persistent organic pollutants (POPs) in sediment cores. As environmental challenges become more complex, NMR provides a precise, reproducible means of identifying both natural and anthropogenic compounds across diverse environmental matrices. NMR spectroscopy plays a vital role in food science by enabling authentication, quality control, and safety analysis of food products. It allows for the identification and quantification of nutrients, additives, and contaminants without destroying the sample. ^1H NMR is particularly useful for profiling metabolites in fruits, vegetables, dairy, and beverages. Recent studies have used NMR to detect food adulteration, such as identifying extra virgin olive oil fraud. Additionally, NMR-based metabolomics aids in assessing freshness and shelf life of food products [9].

5. Future directions of NMR spectroscopy

The future of NMR spectroscopy is marked by transformative innovations expanding their applications across life sciences, chemistry, and materials research. One exciting frontier is in-cell NMR, which enables the observation of biomolecules within living cells, providing unprecedented insights into native protein dynamics and inter-actions. Real-time NMR is increasingly employed for kinetic studies and enzymatic reaction monitoring, allowing researchers to capture transient intermediates and reac-tion pathways as they occur. The integration of quantum computing holds promises for revolutionizing NMR data acquisition and processing, enhancing sensitivity, and reducing computational complexity. Moreover, the fusion of NMR with multi-omics platforms, including genomics, proteomics, and metabolomics, is advancing systems biology by offering a comprehensive, molecular-level view of biological networks and disease mechanisms. Together, these emerging directions will significantly broaden NMR's impact in both fundamental science and applied research.

6. Conclusion

NMR spectroscopy continues to evolve, integrating cutting-edge technologi-cal advances and contributing to transformative breakthroughs across disciplines. From bench research to bedside diagnostics, NMR remains a cornerstone of modern scientific inquiry, and its future applications are poised to expand even further with the support of AI, miniaturization, and interdisciplinary convergence.

Author details

Shagufta Perveen
Department of Bacteriology, University of Wisconsin-Madison, Madison,
WI, United States

*Address all correspondence to: shagufta792000@yahoo.com; sperveen@wisc.edu

IntechOpen

References

[1] Lee JH, Okuno Y, Cavagnero S. Sensitivity enhancement in solution NMR: Emerging ideas and new frontiers. Journal of Magnetic Resonance. 2014;**241**:18-31. DOI: 10.1016/j.jmr.2014.01.005

[2] Pham P, Mandal R, Qi C, Hilty C. Interfacing liquid state hyperpolarization methods with NMR instrumentation. Journal of Magnetic Resonance Open. 2022;**10-11**:100052. DOI: 10.1016/j.jmro.2022.100052

[3] Di Matteo G, Grassi S, Emanuele MC, Scioli G, Brigante FI, Bontempo L, et al. Current applications of benchtop FT-NMR in food science: From quality control to adulteration detection. Food Research International. 2025;**209**:116327. DOI: 10.1016/j.foodres.2025.116327

[4] Shukla VK, Heller GT, Flemming Hansen D. Biomolecular NMR spectroscopy in the era of artificial intelligence. Structure. 2023;**31**(11):1360-1374. DOI: 10.1016/j.str.2023.09.011

[5] Alwahsh M, Nimer RM, Dahabiyeh LA, Hamadneh L, Hasan A, Alejel R, et al. NMR-based metabolomics identification of potential serum biomarkers of disease progression in patients with multiple sclerosis. Scientific Reports. 2024;**14**:14806

[6] Abramson J, Adler J, Dunger J, Evans R, Green T, Pritzel A, et al. Accurate structure prediction of biomolecular interactions with AlphaFold 3. Nature. 2024;**630**:493-500

[7] Capati A, Ijare OB, Bezabeh T. Diagnostic applications of nuclear magnetic resonance-based urinary metabolomics. Magnetic Resonance Insights. 2017;**10**:1-12. DOI: 10.1177/1178623X17694346

[8] Berkson ZJ, Bjorgvinsdottir S, Yakimov A, Gioffre D, Korzynski MD, Barnes AB, et al. Solid-state NMR spectra of protons and quadrupolar nuclei at 28.2 T: Resolving signatures of surface sites with fast magic angle spinning. Journal of the American Chemical Society. 2022;**2**(11):2460-2465

[9] Simpson AJ, Simpson MJ, Soong R. Environmental nuclear magnetic resonance spectroscopy: An overview and a primer. Analytical Chemistry. 2017;**90**(1):628-639. DOI: 10.1021/acs.analchem.7b03241

Chapter 2

Dihedral and Tetrahedral Angles of Five- and Six-Membered Rings Calculated from NMR Data with Three-Sphere Approach

Carmen-Irena Mitan, Emerich Bartha, Petru Filip
and Robert Moriarty

Abstract

3-sphere approach enable calculation of dihedral θ_{HnHn+1} [deg] and tetrahedral angles φ_{Cn} [deg] from NMR data, vicinal coupling constant $^3J_{HH}$ [Hz] and chemical shift δ_{Cn} [deg]. Physical properties used on conformational and configurational analysis. All *cis*, *trans-ee* and *trans-aa* stereochemistry are calculated from three concentric cons with Hopf fibration trigonometric equations that are in agreement with Lie algebra algebraic equations. Exocyclic 3-sphere dihedral angles with right sign and stereochemistry on VISION molecular models give information about the phase angle of the pseudorotation. Dihedral angles are calculated from vicinal angles with four trigonometric equations of ciclyde, torus and inverse of torus Dupin ciclyde, giving the phase angle of the pseudorotation for all possible conformation result from three or four vicinal coupling constants $^3J_{HH}$ [Hz] in case of five- and six-membered rings.

Keywords: 3-sphere, Hopf fibration, Lie algebra, dihedral angles, tetrahedral angles

1. Introduction

3-Sphere, a hypersphere in 4D, under Hopf fibration and Lie group theories, ensure dihedral angles θ_{HnHn+1}[deg] calculation with Hope coordinate and manifold equations from vicinal coupling constant $^3J_{HH}$ [Hz] and/or chemical shift δ [ppm], and vicinal angles from $^3J_{HH}$ [Hz] with Lie algebra equations, a method studied around 1998 by Dr. Ing. E. Bartha and gives to scientific people in 2021 [1–3]:

$$^3J_{HH} = \frac{\sqrt{\phi + \cos 2\theta_{HnHn+1}}}{2} \qquad (1)$$

where $^3J_{HH}$ [Hz] – vicinal coupling constant, ϕ [deg] – vicinal angle, θ_{HnHn+1}[deg] – dihedral angle.

Equation 1 demonstrates a strong correlation between the dihedral angle, the vicinal angle, and the vicinal coupling constant [4]

3-Manifold locally reassembles Euclidean 3-dimensional space and can be considered the shape of the universe. Topological at a smaller distance relative to the observer, a sphere is visualized as a plane surface, that is, earth is a sphere from space, since at a higher distance, two parallel lines give a triangle, and a Greek column only a concave surface, neither a sphere nor a triangle. *Nash theorem*, isometrically embedded in Euclidean space R^n are Riemman manifolds. *Gauss-Bonnet theorem*, the integral of the Gauss curvature on compact two-dimensional Riemannian manifold is equal to $2\pi\lambda(M)$, with $\lambda(M)$ Euler characteristic of M. *Henri Poincaré theorem concern space*, ordinary 3-dimensional' space is connected finite in size, and locks any boundary. Each loop in the space is continuously heightened to a point but requires a 3-dimensional sphere. Each distinct point on 2-sphere comes from a distinct circle on 3-sphere, continuous function from 3-sphere into 2-sphere. Hopf fibration, visualized as a stereographic projection of S^3 to R^3, and R^3 to a ball, gives cyclide – a torus from inverse image of latitude circle in S^2 under the fiber map, with Villarceau circles as fibers. A collection of fibers over circles in S^2 is a torus ($S^1 x S^1$), and every pair of tori are linked once. An example gives constantly about Poincaré conjecture is coffee cup with its torus and cylindrical coordinates, limit case of conic. The geodesics on the unroll cone cut along the meridian is a straight line in plane, a way to translate in 2D. Dodecahedron tessellation in H^3 gives the possibility to see inside of hyperbolic 3-manifold:

$$S^n(r) = \{x \in Rn + 1 : IIXII = r\}, r^2 = \sum_{i=1}^{n+1} (xi - ci)^2 \qquad (2)$$

The set of points in (n + 1)-space $(X_1, X_2, ..., X_{n+1})$ that define n-sphere, where $C = (c_1, c_2, ..., c_{n+1})$ is a center point, and r is the radius. R^4 is a I. complex C^2 is a Hilbert space, or an inner product, or II. quaternions (H) described the hypersphere S^3 embedded in R^4:

I. Complex C^2:

$$S^3 = \{(Z_1, Z_2) \epsilon C^2 : IZ_1I^2 + IZ_2I^2 = 1\} \qquad (3)$$

where $Z_1 = e^{i\zeta 1}sin\theta, Z_2 = e^{i\zeta 2}cos\theta, \phi\epsilon [0, 2\pi], \theta\epsilon [0, \pi/2]$

Rotation around the origin in 4D space has two invariants' planes orthogonal to each other, intersected at origin, and are rotated by two independent angles (ζ^1 and ζ^2).

Hopf coordinates expressed in R^4:

$$X_0 = cos\zeta^1sin\theta, X_1 = sin\zeta^1sin\theta \qquad (4a,b)$$

$$X_2 = cos\zeta^2sin\theta, X_3 = sin\zeta^2sin\theta \qquad (5a,b)$$

II. Quaternions:

Any points on the 3-sphere equivalent to a quaternion is equivalent to a particular Cartesian coordinate rotation in three dimensions. All points in the 2-sphere give a possible set of quaternions with possible rotations. For any point P in S^3 the preimage set $h^{-1}(P)$ is a circle in S^2. Hamilton, in 1843, discovered quaternion, algebra of rotations in R^3 using 4-tuples, in other words, two-dimensional rotation by complex number C. Heinz Hopf, in 1931, claimed that unit sphere in complex coordinate space C^{n+1} fibers over complex projective space CP^n with circles as fibers:

$$S^3 = \{q\epsilon H : IIqII = 1\} \qquad (6)$$

Euler's analog:

$$q = e^{\tau\psi} = \cos\psi + \tau\sin\psi, \tau^2 = -1 \tag{7a}$$

$$\tau = (\cos\theta)xi + (\sin\theta\cos\zeta)xj + (\sin\theta\sin\zeta)xk \tag{7b}$$

$$q = e^{\tau\psi} = X_0 + X_1 i + X_2 j + X_3 k \tag{7c}$$

where q describes spatial rotations, a rotation about τ through an angle of 2ψ; τ imaginaries quaternion.

The quaternion group (Cycle graph) is a non-abelian group of order eight, isomorphic to a certain eight elements, and has the same order as the dihedral group DiH (Cayley graph) but with a different structure.

Hopf fibration is a projection between the hypersphere S^3 of the quaternions in 4D space and the unit sphere S^2 in 3D space. Since complex numbers, C described rotations in 2D, the quaternions C^2 described rotations in both 4D and 3D. Points in R^4 are three-sphere in S^3 [5–8].

2. 3-Sphere and *cis-*, *trans-ee*, and *trans-aa* stereochemistry

cis- and *trans*-Stereochemistry in higher dimension space can be visualized as three concentric cons with six dihedral angles, three pairs of *cis-trans* angles, and two characteristics angles ϕ_1 and ϕ_2, namely vicinal angles (**Figure 1**).

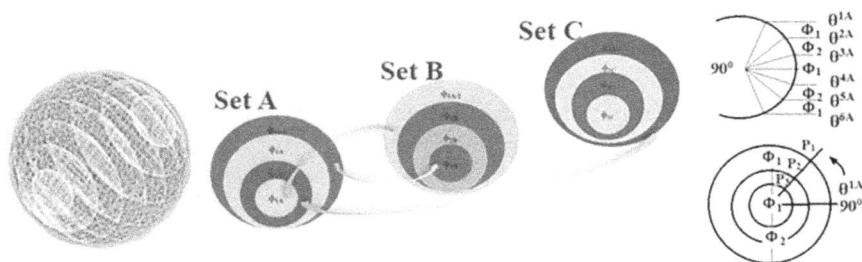

Figure 1.
Three-sphere: complex C^2 and cis- *and* trans-*stereochemistry.*

I. *Complex C^2*: Calculation vicinal coupling constant $^3J_{HH}$ [Hz] from vicinal angle ϕ [deg] (8), (9) [4]:

$$^3J_{HH} = \frac{\sqrt{\phi^{cis,trans-ee}}}{2} \tag{8}$$

$$^3J_{HH} = \sqrt{\phi^{trans-aa}} \tag{9}$$

where $^3J_{HH}$ – vicinal coupling constant [Hz],
ϕ - vicinal angle [deg].

II. *Quaternions*: Spherical representation in higher dimensional space with cyclic graph (**Figure 2**) [4]:

The vicinal coupling constant $^3J_{HH}$ [Hz] can be calculated from the differences between two dihedral angles under *cis-trans* relationships with Hückel I and II equations.

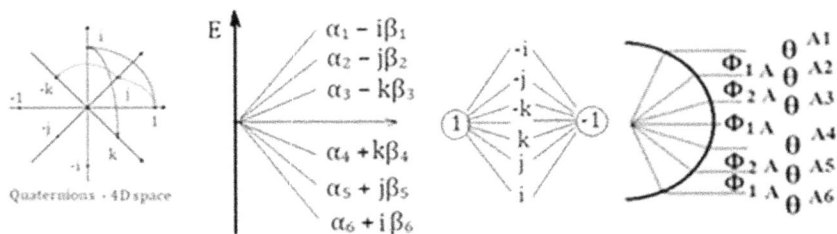

Figure 2.
Quaternions multiplications (H) and cis, trans *stereochemistry.*

Hückel I (10) and Hückel II (11) equations:

$$E = \alpha + 2mix\beta \tag{10}$$

$$E = \alpha + mixI2\beta I \tag{11}$$

where $m_j = \cos(2xjx180/n)$, i, j, k, $-i$, $-j$, $-k$ – quaternionic, analog to j = 1–7 angular momentum.

Vicinal coupling constant calculation $^3J_{HH}$ [Hz] (12), (13):

$$^3J_{HH} = \sqrt[n]{\frac{\theta I - \theta II}{mi}} \tag{12}$$

$$^3J_{HH} = \sqrt[n]{\frac{\frac{\theta I - \theta II}{mi}}{2}} \tag{13}$$

where *cis* n = 4, *trans* n = 2, I, II – pairs of *cis-trans* angles: 1, 6; 2, 5; 3, 4.

Quaternionic and octonionic Hopf fibration are proposed for constants coupling prediction [4, 9]. The distribution of the m_i values on the three concentric cones reveals that S^{15} is isomorphic to the Lie algebra E_8. Vedic Nuclear Physics is a parallel construction between Lie Algebra E_8 and Hopf fibration [10, 11].

2.1 Building units

Projection of a 3-sphere in 3D (**Figure 3**), three concentric cons with two characteristic angles are used for building two sets of angles with *cis-*, *trans-ee*, *trans-aa* stereochemistry, where characteristic angles become angles of set B, $\phi_1/2$ of set A is first angle of set B and $\phi_1/2$ of set B is first angle of set A [12].

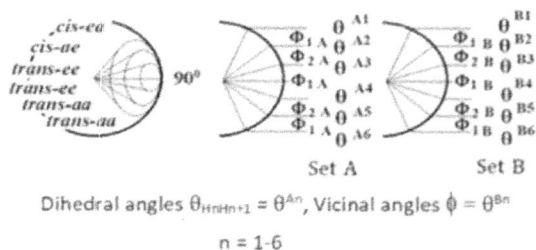

Dihedral angles $\theta_{HnHn+1} = \theta^{An}$, Vicinal angles $\phi = \theta^{Bn}$

n = 1-6

Figure 3.
Three-sphere in 3D projection and cis, trans-ee, trans-aa *stereochemistry.*

One unit with seven sets of angles results from relationships between angles. The first angle can be calculated from the *sin* function, resulting in unit U with angles higher than 5[deg], or from the *tan* function, resulting in unit S with angles smaller than 5[deg]. In function of the values of the first angles on seven sets unit are remarkably three pairs of sets: A, B, C; D, E, A; F, G, B with following successions U-S-U, U-U-S, S-U-S [12].

Seven sets angles: θ^{CnAi}, θ^{TnAi}: C = U, S; T = U, S.

UnN1 > 5[deg]; Un = UnAi, UnBi, UnCi, UnDi, UnEi, UnFi, UnGi, n = 1–6.

SnN1 < 5[deg]; Un = SnAi, SnBi, SnCi, SnDi, SnEi, SnFi, SnGi, n = 1–6.

Six sets of angles on two units, two pairs of three sets of angles A, B, C under U and S unit rule, with the possibility to transform from U to S (14) and S to U (15):

UnN1 > 5[deg]; Un = UnAi, UnBi, UnCi, n = 1-6.

SnN1 < 5[deg]; Un = SnAi, SnBi, SnCi, n = 1-6

$$\theta^{S1A1} = \left(\phi_2^{U1A} - \frac{\phi_1^{U1A}}{2} \right) - \text{Transformation U to S} \tag{14}$$

$$\theta^{U1B1} = \frac{\phi_1^A}{2} = \frac{\phi_2^B}{2} = \frac{\frac{60+/-\theta^{S1A1}}{1.5}}{2} - \text{Transformation S to U} \tag{15}$$

Sets A and B are in close relationship with set C through ϕ_2^A, set C a way for building second unit U_2 or S_2. The unit can be increased at seven transforming first angles of set A and B in ϕ_2 of D and F. Between units U and S, values of angles are almost equal with values obtained from trigonometric function *sin* and *tan* from carbon or/and proton chemical shift (14), (15). Probably six sets of angles on two units lost utility once the seven sets of angles were ensured for the example in work relationship U to S (**Figure 4**).

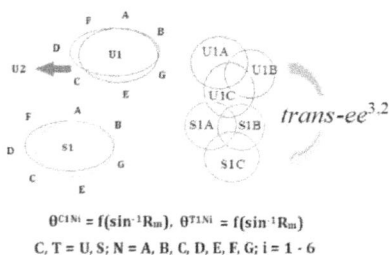

$$\theta^{C1Ni} = f(\sin^{-1}R_m), \quad \theta^{T1Ni} = f(\sin^{-1}R_m)$$
$$C, T = U, S; \; N = A, B, C, D, E, F, G; \; i = 1 - 6$$

Figure 4.
Seven sets of angles and six sets of angles on units U and S. Unit S build through algebraic equation for trans-ee^{3,2}
stereochemistry.

Relationships between tetrahedral angles φ_{Cn} [deg] and dihedral angles θ_{HnHn+1}[deg] of five- and six-membered rings can be established on units built from *sin* and *tan* function (16), (17). I. Unit build from vicinal angle ϕ [deg] calculated from coupling constant $^3J_{HH}$ [Hz]. II. From carbon chemical shift or/and proton chemical shift δ_{Cn} [ppm] under manifold rule, that is, Euler-conic (18), are calculated four angles used for building two units with seven sets angles. On the examples already published, relationships between dihedral θ_{HnHn+1}[deg] and tetrahedral angles φ_{Cn} [deg] under trigonometric function can be found on five units, in fact, four (A, B, D, E), set C is just a set of transition. Because a few examples are observed relationships between dihedral θ_{HnHn+1} [deg] and tetrahedral φ_{Cn} [deg] angles on sets A, B, F, and G, it probably is better to calculate seven sets [12]:

$$\cos\theta^A = R_m = \sin\theta^B \qquad (16)$$

$$\tan\frac{1}{\theta^A} = R_m = \tan\theta^B \qquad (17)$$

$$R_m = \left(\frac{\nu x 4\pi x 10^{-3}}{\gamma}\right) \text{[gauss} \times 10] \qquad (18)$$

where θ^A, θ^B – angles of sets A and B [deg],
ν – frequency [Hz], $\nu = \delta x \omega_L$,
ω_L – Larmor frequency [^{13}C: 75 MHz],
δ_{Cn} – chemical shifts [ppm],
γ – gyromagnetic ratio: ^{13}C: $\gamma = 10.71$ [MHzxT^{-1}] = 6.7[10^7xradxT^{-1}xs^{-1}];

$$B_0 = \left(\frac{2\pi x\, \omega L}{\gamma}\right) \left[\text{MHzxT}^{-1}\right] = \frac{2\pi x\, 75}{10.71} = 7.0028[\text{T}], \qquad (19)$$

where B_0 – magnetic field [gauss],
H - field intensity,
$B = \mu H$ [oersted],
μ – magnetic permeability,
$1[\text{T}] = 10^4[\text{gauss}] = 10^3[\text{A/m}]$,
$1[\text{oersted}] = 10^3/4\pi$ [A/m].

2.2 Trigonometric equations: Hopf fibration

Heinz Hopf describes a 3-sphere in terms of a circle and an ordinary sphere. Hopf fibration is equivalent to the fiber bundle structure of the Dirac monopole. In quantum mechanics, the Riemann sphere known as Block sphere describes topological structure of a quantum mechanical two-level system or qubit. Generalization of the double qubits to single qubits of Bloch sphere, in other words higher dimension of Hopf fibration generalized to real Hopf fibration [13].

Four bundles in which the base space and fiber space are all sphere is known as Hopf fibration:

One-sphere: real Hopf fibration $S^0 \to S^1 \to S^1$,

3-sphere: complex Hopf fibration $S^1 \to S^3 \to S^2$,

Seven-sphere: quaternionic Hopf fibration $S^3 \to S^7 \to S^4$,

Fifteen-sphere: octonionic Hopf fibration $S^7 \to S^{15} \to S^8$.

Higher dimension Hopf fibration under stereographic projection, CP^1 identified as S^2, HP^1 with S^4 and OP^1 with S^8 [14]:

Complex projective space: $S^1 \to S^{2n+1} \xrightarrow{\pi} CP^n$, 2n real dimension,

Quaternionic projective space: $S^3 \to S^{4n+1} \xrightarrow{\pi} HP^n$, 4n real dimension,

Octonionic projective space: $S^7 \to S^{15} \xrightarrow{\pi} OP^n$, 8n real dimension.

Spheres S^1, S^3, and S^7 are parallelizable, corresponding to unit norm sets of complex numbers. S^1 and S^3 are Lie groups, and octonionic Hopf fibration is divided with Moufang loop [15] or Moufang quasigroup [16] structure on seven-sphere S7.

The octonionic analogy of the quaternionic group Sp(2).Sp(1)cSO(8) is Spin(9) cSO(16). The Riemannian holonomies Spin(9) is only possible on manifold (M^{16}) that are either flat or locally isometric to OP^2 or on hyperbolic Cayley plane OH^2 [17].

Single qubit Hilbert space is a sphere S^3 in 3D, and two qubits Hilbert space is a sphere S^7 in 7D, [13] fibration of seven-sphere over four-sphere with fiber of 3-sphere, in case of octonions with fiber seven-sphere. Coordinates between two sets angles: R^4 – complex Hopf fibration (17a,b), (18a,b), trigonometric relationships between angles θ^A and θ^B of sets A and B (17c) and *vice versa* between angles θ^B and θ^A of sets B and A (18c):

$$X_0 = \cos\theta^B \sin\theta^A, X_1 = \sin\theta^B \sin\theta^B, \qquad (17a,b^*)$$

*previously presented as Eq. (4a,b)

$$X_2 = \cos\theta^A \sin\theta^B, X_3 = \sin\theta^A \sin\theta^A. \qquad (18a,b^*)$$

*previously presented as Eq. (5a,b)
Where: if $\theta^b = \zeta^1$, $\theta = \theta^A$ (4a,b), else $\theta^A = \zeta^2$, $\theta = \theta^B$ (5a,b)

$$\theta^B = \cos^{-1}\sin\theta^A, \qquad (17c^*)$$

*previously presented as Eq. (4c)

$$\theta^A = \cos^{-1}\sin\theta^B. \qquad (18c^*)$$

*previously presented as Eq. (5c)
Trigonometric equations between two sets of angles are obtained from Euller tessellation, analog to Hopf coordinates, embedded S^3 in C^2 (θ, ϕ_1, ϕ_2). Relationships between sets A, B, and C can be demonstrated with quaternions (H), Euler's analog equation. Quaternions are points on R^4, and for any point P in S^3 the preimage s. $h^{-1}(P)$ is a circle in S^2. A circle of latitude gives a torus (T^2 in R^3); more circles of latitude give tori [18].

2.3 Stereochemistry *cis*, *trans* and Hopf coordinates

Dihedral angles with positive signs on the east side of the hypersphere are in R^{16}, admitting the coordinates X_6, X_7 and X_{10}, X_{11} for *trans-ee* stereochemistry in units U and S; otherwise, they are in R^{12} in strict application rule. Two sets of angles with 24 angles with negative and positive signs on the circle representation have two sets of equations R^{16}, a total of R^{32}. Considering coordinates expressed in R^4, rotation around the origin in 4D space with ζ^1 angle of set A and ζ^2 angle of set B (4a,b, 5a,b), Eqs. (19a,b)–(26a,b) ensures real Hopf fibration, calculation of the dihedral angles $\theta_{HnHn+1} = \theta^B$ from vicinal angles $\phi = \theta^A = \zeta^1$. Calculation of the dihedral angles $\theta_{HnHn+1} = \theta^A$ from vicinal angles $\phi = \theta^B = \zeta^2$ giving second real Hopf fibration from complex Hopf fibration. In both cases is quantified the relationship with set C, important point for building other units [12]:
If $\phi = \theta^A$, $\theta_{HnHn+1} = \theta^B$:

$$trans - aa^{6,1} : X_0 = \cos\theta^{U1B3}\sin\theta^{U1A6}, X_1 = \sin\theta^{U1B3}\sin\theta^{U1C5}/2 \qquad (19a,b)$$

$$trans - aa^{5,2} : X_2 = \cos\theta^{U1B2}\sin\theta^{U1A5}, X_3 = \sin\theta^{U1B2}\sin\theta^{U1C4}/2 \qquad (20a,b)$$

$$trans - ee^{4,1} : X_4 = \cos\theta^{U1B1}\sin\theta^{U1A4}, X_5 = \sin\theta^{U1B1}\sin\theta^{U1C1}/2 \qquad (21a,b)$$

$$trans - ee^{3,2} : X_6 = \cos\theta^{U1B1}\sin\theta^{U1A3}, X_7 = \sin\theta^{U1B1}\sin\theta^{U1C1}/2 \qquad (22a,b)$$

$$trans - ee^{3,2} : X_8 = \sin\theta^{S1B1}\cos\theta^{S1A3}, X_9 = \sin\theta^{S1B1}\sin\theta^{S1C1}/2 \qquad (23a,b)$$

$$trans - ee^{4,1} : X_{10} = \sin\theta^{S1B1}\cos\theta^{S1A4}, X_{11} = \sin\theta^{S1B1}\sin\theta^{S1C1}/2 \qquad (24a,b)$$

$$cis - ae/ea^{1,6} : X_{12} = \cos\theta^{U1B3}\sin\theta^{U1A1}, X_{13} = \sin\theta^{U1B3}\sin\theta^{U1C5}/2 \qquad (25a,b)$$

$$cis - ae/ea^{2,5} : X_{14} = \cos\theta^{U1B2}\sin\theta^{U1A2}, X_{15} = \sin\theta^{U1B2}\sin\theta^{U1C4}/2 \qquad (26a,b)$$

Four sets of Eqs. (27)–(29), (30)–(32), (33)–(35), (36)–(40) are established between three sets of angles drawn on 2D circles (**Figure 5**), covering all relationships between angles A and B, vicinal – dihedral, between the east and west side of the circle with all characteristic angles for north and south. Relationships between sets B and A giving R^{64}. Three sets of angles are easy to simulate inside of the hypersphere and are represented as a circle in 2D. Representation of three sets of angles inside the sphere has at base coordinates of successive rotation of Poincaré rotator, with two main coordinates (1) rose and (2) torus [19].

Eq. 33: $\cos^{-1}\sin\text{-}\theta^{A1} = \theta^{B4}$		Eq. 27: $\cos^{-1}\sin\theta^{A1} = \theta^{B3*}$
Eq. 34: $\cos^{-1}\sin\text{-}\theta^{A2} = \theta^{B5}$	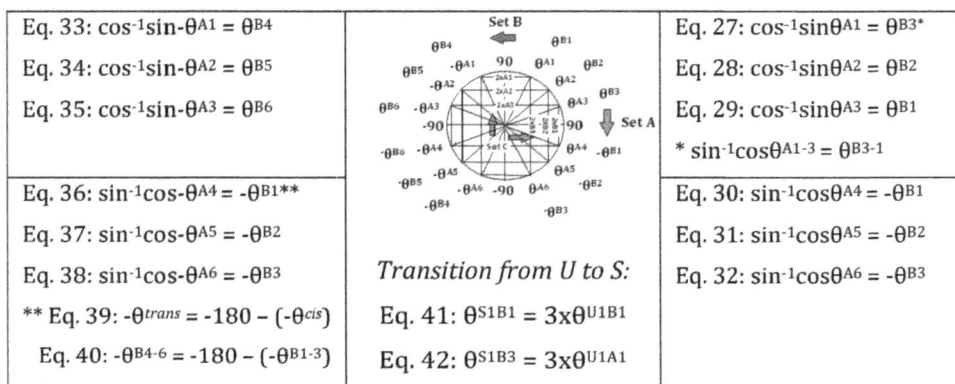	Eq. 28: $\cos^{-1}\sin\theta^{A2} = \theta^{B2}$
Eq. 35: $\cos^{-1}\sin\text{-}\theta^{A3} = \theta^{B6}$		Eq. 29: $\cos^{-1}\sin\theta^{A3} = \theta^{B1}$
		* $\sin^{-1}\cos\theta^{A1\text{-}3} = \theta^{B3\text{-}1}$
Eq. 36: $\sin^{-1}\cos\text{-}\theta^{A4} = -\theta^{B1}$**		Eq. 30: $\sin^{-1}\cos\theta^{A4} = -\theta^{B1}$
Eq. 37: $\sin^{-1}\cos\text{-}\theta^{A5} = -\theta^{B2}$		Eq. 31: $\sin^{-1}\cos\theta^{A5} = -\theta^{B2}$
Eq. 38: $\sin^{-1}\cos\text{-}\theta^{A6} = -\theta^{B3}$	*Transition from U to S:*	Eq. 32: $\sin^{-1}\cos\theta^{A6} = -\theta^{B3}$
** Eq. 39: $-\theta^{trans} = -180 - (-\theta^{cis})$	Eq. 41: $\theta^{S1B1} = 3x\theta^{U1B1}$	
Eq. 40: $-\theta^{B4\text{-}6} = -180 - (-\theta^{B1\text{-}3})$	Eq. 42: $\theta^{S1B3} = 3x\theta^{U1A1}$	

Figure 5.
Four sets of equations between sets A and B.

2.4 3-sphere dihedral angles

Considering angles θ^A and θ^B, dihedral angle θ_{HnHn+1}[deg] and vicinal angle ϕ [deg] or *vice versa*, with angular velocity representation (**Figure 3**) are point out the relationship between angles of sets A, B, and C on triangles. Wonderful relationship results with differential topology between angle and inverse of angle on two sets: $\theta^A x (\theta^B)^{-1}$ or $\theta^B x (\theta^A)^{-1}$, known as homotopy in mathematics, and characteristic for wave behavior (**Figure 6**).

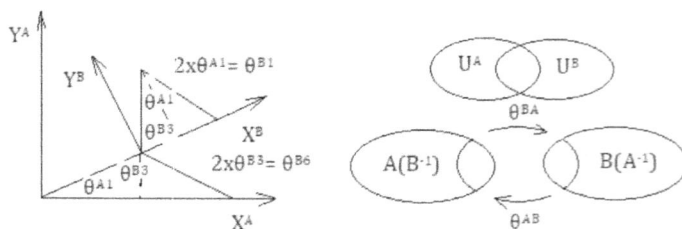

Figure 6.
Angular velocity in R^4exemplified for tran-aa^{6,1} *stereochemistry. Differential topology, the relationship between angles of set A and B: that is, between θ^{A1}, θ^{B3} for cis^{6,1} stereochemistry and between θ^{A2}, θ^{B2} for cis^{5,2} stereochemistry.*

Six angles on set A with *cis*, *trans-ee*, and *trans-aa* stereochemistry ensure $\mathbf{R^{16}}$: octonionic *Hopf fibration*, quaternionic multiplication, in fact, $\mathbf{R^{32}}$ for all stereochemistry with positive and negative signs [12].

From the NMR data point of view, trigonometric relationship between vicinal and dihedral angles is just real Hopf fibration: $\mathbf{S^0} \rightarrow \mathbf{S^1} \rightarrow \mathbf{S^1}$. Equation of torus (43) where dihedral angle θ_{HnHn+1} [deg] can be calculated from vicinal angle ϕ [deg] results from vicinal coupling constant $^3J_{HH}$ [Hz] for *cis*-stereochemistry, and Eq. (44) for transformation *cis* to corresponding *trans* angle [1, 20].

Inverse of torus is Dupin cyclide (**Figure 7**), both having four characteristic circles, toroidal, poloidal, and two Villarceau circles, which results at the intersection of a torus with a bitangent plane. The tangential space of Dupin cyclide and *sin* of *torus* on 3D Eqs. (45)-(47), gives other two dihedral angles with different signs and stereochemistry from the same vicinal coupling constant $^3J_{HH}$ [Hz] [21].

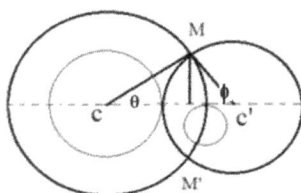

Figure 7.
Torus and invers of torus Dupin Ciclide.

$$\theta_{HnHn+1} = \sin^{-1}\cos\phi \qquad (43)$$

$$\theta_{HnHn+1} = \cos^{-1}\sin\phi \qquad (44)$$

$$\theta_{HnHn+1} = \tan^{-1}\sin - \phi \qquad (45)$$

$$\theta_{HnHn+1} = \sin^{-1}\frac{1}{\tan - \phi} \qquad (46)$$

$$\theta_{HnHn+1} = \sin^{-1}\tan - \phi \qquad (47)$$

Dihedral angles can be calculated also from carbon and/or proton chemical shift using manifold equations: conic section, Euler-conic, rectangle, Villarceau circles, and polar equation, followed by building units and choosing dihedral angle with values almost equal to the recorded vicinal angle. The calculated vicinal coupling constant must be almost equal to the recorded one, with differences of a few degrees corresponding to other sets of angles [12].

The vicinal angle ϕ [deg] can be calculated from the vicinal coupling constant $^3J_{HH}$ [Hz] for values between 1 and 6[Hz] for *cis*, *trans-ee* (48) and between 6 and 13[Hz] for *trans-aa* (49), [20, 22] and *vice versa trans-aa* (48) and *trans-ee* (49) [23]:

$$\phi = \left(2x^3 J_{HH}\right)^2 \qquad (48)$$

$$\phi = \left(^3 J_{HH}\right) \qquad (49)$$

2.5 Algebraic equations on *cis*, *trans* stereochemistry

Algebraic equations for all stereochemistry are demonstrated with the A2 root system Lie group in **Figure 8**. Six vectors in two-dimensional Euclidean space R^2, namely roots, with equilateral triangles of the Weyl group [11, 24]:

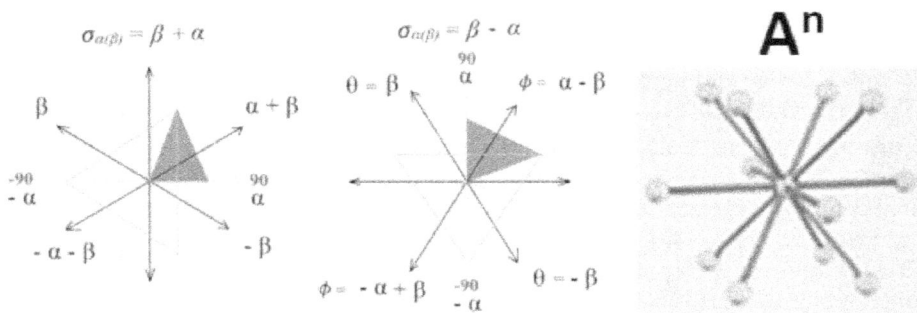

Figure 8.
Trans/cis-*stereochemistry and A2 root system* – trans: $90 + \theta^B = \theta^A$; cis: $90 - \theta^A = \theta^B$.

$$\sigma_\alpha(\beta) = \alpha{-}\beta^D \qquad (50)$$

$$\sigma_\alpha(\beta) = \alpha{-}\beta^S \qquad (51)$$

where all the negative angles on the left (S) and all positive angles on the right (D):

$$cis/trans - aa : \sigma_\alpha(\beta) = \beta{-}\alpha \qquad (52)$$

where: β^D or $\beta^S = \theta^{nN}$

$$trans - ee^{1,4} : \sigma_\alpha(\beta) = \alpha + \beta^D \qquad (53)$$

where $\beta^D = \phi_N$.

Dihedral angles θ_{HnHn+1} [deg] are under the 30/60/120 [deg] rule of equilateral triangle on sets with six angles covering all stereochemistry, values which fit well on root system Lie group, Weyl group of the A2 root system. Algebraic equations are in agreement with trigonometric equations except *trans-ee* stereochemistry. The vicinal coupling constant $^3J_{HH}$ [Hz] or vicinal angle ϕ [deg] can be calculated from characteristic angles of set A: $\phi_1{}^A/2$: the difference between 90 [deg] and θ^{A3} [deg], $\phi_2{}^A$: the difference between θ^{A3} [deg] and θ^{A2} [deg], at list $\phi_1{}^A$: the difference between θ^{A2} [deg] and θ^{A1} [deg]. The characteristics angles of set A $\phi_1{}^A/2$ and $\phi_2{}^A$ are angles of set B. Algebraic equations are established for all stereochemistry *cis–/trans-aa*[6,1] (54), *cis–/trans-aa*[5,2] (55), *trans-ee*[3,2] (56), and *trans-ee*[4,1] (57) with positive sign and negative sign *cis–/trans-aa*[6,1] (58), *cis–/trans-aa*[5,2] (59), *trans-ee*[3,2] (60), and *trans-ee*[4,1] (61).

Algebraic equations for dihedral angles with positive sign [12]: $^3J_{HH} = f(\phi_1/2, \phi_2)$

$$cis/trans - aa^{6,1} : {}^3J_{HH}{}^{6,1} = (\phi^{6,1})^{1/2}/m = (60 + \phi_1{}^A/2)^{1/2}/m \qquad (54)$$

$$cis/trans - aa^{5,2} : {}^3J_{HH}{}^{5,2} = (\phi^{5,2})^{1/2}/m = (\phi_2{}^A + \phi_1{}^A/2)^{1/2}/m \qquad (55)$$

$$trans - ee^{3,2} : {}^3J_{HH}{}^{3,2} = (\phi^{3,2})^{1/2}/m = (\phi_2{}^A - \phi_1{}^A/2)^{1/2}/m \qquad (56)$$

$$cis/trans - ee^{4,1} : {}^3J_{HH}{}^{4,1} = (\phi^{4,1})^{1/2}/m = (\phi_1{}^A)^{1/2}/m \qquad (57)$$

m = 1 *trans-aa*, m = 2 *cis-*, *trans-ee*
Algebraic equations for dihedral angles with negative sign: $^3J_{HH} = f(\phi_1/2, \phi_2)$

$$cis/trans - aa^{6,1} : {}^3J_{HH}{}^{6,1} = (\phi^{6,1})^{1/2}/m = (120 - \phi_1{}^A/2)^{1/2}/m \qquad (58)$$

$$cis/trans - aa^{5,2} : \, {}^{3}J_{HH}{}^{5,2} = \left(\phi^{5,2}\right)^{1/2}/m = \left(120 + \phi_1{}^A/2\right)^{1/2}/m \quad (59)$$

$$trans - ee^{3,2} : {}^{3}J_{HH}{}^{3,2} = \left(\phi^{3,2}\right)^{1/2}/m = \left[180 - \left(120 - \phi_2{}^A - \phi_1{}^A/2\right)\right]^{1/2}/m \quad (60)$$

$$cis/trans - ee^{4,1} : \, {}^{3}J_{HH}{}^{4,1} = \left(\phi^{4,1}\right)^{1/2}/m = \left[180 - \left(120 + \phi_2{}^A + \phi_1{}^A/2\right)\right]^{1/2}/m \quad (61)$$

m = 1 *trans-aa*, m = 2 *cis-, trans-ee.*

2.6 Algebraic and trigonometric equations. Lie algebra and Hopf fibration

The topology of SO(4) is compatible with Lie Group SO(3)xSpin(3) = SO(3)xSU (2), topology of P^3xS^3. Trigonometric equations for *trans-ee*^{3,2} and *trans-ee*^{4,1} give two values from the same vicinal angle ϕ [deg], positive for first and negative ϕ [deg] for second. Algebraic Eq. (56) for *trans-ee*^{3,2} stereochemistry on unit U gives first angle of set A or B with values smaller as 5 [deg] of unit S, that is, θ^{S1B1} 2.84[Hz] (**Table 1**). For *trans-ee* stereochemistry are expected smaller values of vicinal coupling constant $^{3}J_{HH}$ [Hz] on unit S. As the main observation, on unit S *trans-ee* angle coming from 3, 2 stereochemistry of unit U, as demonstrated with algebraic Eq. (56), is again under the trigonometric equations with uncertain attribution of stereochemistry.

θ^{C1Nn} [deg]	Set A	Set B	θ^{C1Nn} [deg]
θ^{U1A1}	20.94	9.05	θ^{U1B1}
θ^{U1A2}	39.05	50.94	θ^{U1B2}
θ^{U1A3}	80.94	69.05	θ^{U1B3}
θ^{U1A4}	99.05	110.94	θ^{U1B4}
θ^{U1A5}	140.94	129.05	θ^{U1B5}
θ^{U1A6}	159.05	170.94	θ^{U1B6}
ϕ_2	41.89	18.10	ϕ_2
$\phi_1/2$	9.05	20.94	$\phi_1/2$
$\theta^{S1B1} = 3x\theta^{U1B1}$		$\theta^{S1B3} = 3x\theta^{U1A1}$	
27.84	θ^{S1A1}	θ^{S1B1}	2.84
32.84	θ^{S1A2}	θ^{S1B2}	57.15
87.15	θ^{S1A3}	θ^{S1B3}	62.84
92.84	θ^{S1A4}	θ^{S1B4}	117.15
147.15	θ^{S1A5}	θ^{S1B5}	122.84
152.84	θ^{S1A6}	θ^{S1B6}	177.15
54.30	ϕ_2	ϕ_2	5.64
2.84	$\phi_1/2$	$\phi_1/2$	27.15
S to U: if $\phi_2 < \phi_1/2$: $\theta^{U1A1} = [(60 + \theta^{S1B1})/1.5]/2$.			

Table 1.
Four sets of angles on two units (U and S) for a vicinal coupling constant of 1.0[Hz] and 68.0[ppm].

Positive *gauche* angle results from vicinal coupling constant of 2.6 and 2.7 [deg] as calculated for *cis*-stereochemistry, and negative *gauche* angle results from *trans* stereochemistry rule and a vicinal coupling constant of 12.3 [Hz], in both cases calculated with Eq. (43). Unit S results from Eq. (45) at 2.7[Hz] and from Eq. (47) at 2.5[Hz], with angles from which can be calculated *trans-ee*^{4,1} from *cis* under 120[deg] rule. At 3.1[Hz], results from Eq. (45), a *cis* angle of −31.86[deg] corresponding to −88.13 [deg] with *trans-ee*^{3,2} stereochemistry. Trigonometric Eqs. (46)–(48) give negative values for *cis⁻*stereochemistry for vicinal coupling constant of 0–1.1 [Hz], negative (46) and positive (48) for vicinal coupling constant higher as 13.2 [Hz], and the transformation on *trans-ee* stereochemistry under +/−120[deg] rule is uncertainly:

Other observation, in contrast with known rule +/−120 [deg] between *cis-* and *trans-* stereochemistry, from trigonometric equations results from the same vicinal angle *cis-* and *trans-aa* dihedral angles under 180 [deg] rule, and *cis-*, *trans-ee* dihedral angles under 120 rule (**Table 2**) [25].

Stereochemistry	*cis, trans-aa*[6,1] 180[deg] rule	*cis, trans-aa*[5,2] 180[deg] rule	*cis, trans-ee* 120[deg] rule
cis[6,1]	$\theta^{A1}\,\theta^{B1}$	$\theta^{A1}\,\theta^{B1}$	$\theta^{A1}\,\theta^{B1} = \phi$
cis[5,2]	$\theta^{A2}\,\theta^{B2}$	$\theta^{A2}\,\theta^{B2} = \phi$	$\theta^{A2}\,\theta^{B2}$
trans-ee[3,2]	$\theta^{A3}\,\theta^{B3} = \phi$	$\theta^{A3}\,\theta^{B3}$	$\theta^{A3}\,\theta^{B3}$
trans-ee[4,1]	$\theta^{A4}\,\theta^{B4}$	$\theta^{A4}\,\theta^{B4}$	$\theta^{A4}\,\theta^{B4}$
trans-ee[5,2]	$\theta^{A5}\,\theta^{B5}$	$\theta^{A5}\,\theta^{B5}$	$\theta^{A5}\,\theta^{B5}$
trans-aa[6,1]	$\theta^{A6}\,\theta^{B6}$	$\theta^{A6}\,\theta^{B6}$	$\theta^{A6}\,\theta^{B6} = \phi$
$\cos^{-1}\sin\phi = \theta^{HnHn + 1}$	$\cos^{-1}\sin\phi = \theta^{A1}$ $\cos^{-1}\sin(-\phi) = \theta^{A6}$ $180 - \theta^{A6} = \theta^{A1}$	$\cos^{-1}\sin\phi = \theta^{A2}$ $\cos^{-1}\sin(-\phi) = \theta^{A5}$ $180 - \theta^{A5} = \theta^{A2}$	$\cos^{-1}\sin\phi = \theta^{A3}$ $\cos^{-1}\sin(-\phi) = \theta^{A4}$ $\theta^{A2} = \theta^{A3}$-120 $\theta^{A1} = \theta^{A4}$-120

ϕ – *vicinal angle [deg],* $\theta^{HnHn + 1}$ - *dihedral angle [deg].*

Table 2.
Three-phere torsional angles for trans-ee *and* trans-aa *stereochemistry.*

2.7 Four-dimensional (4D) space: Translation from sphere to torus

The transformation from U to S is probably the best example for translation from sphere to torus (**Figure 9**); in other words, a hole inside of the sphere gives tori. "Stereogenic projection of 3-sphere mapped in Euclidean 3-space, the inverse of a circle of latitude on S^2 under the fiber map is a torus, and the fibers themselves are

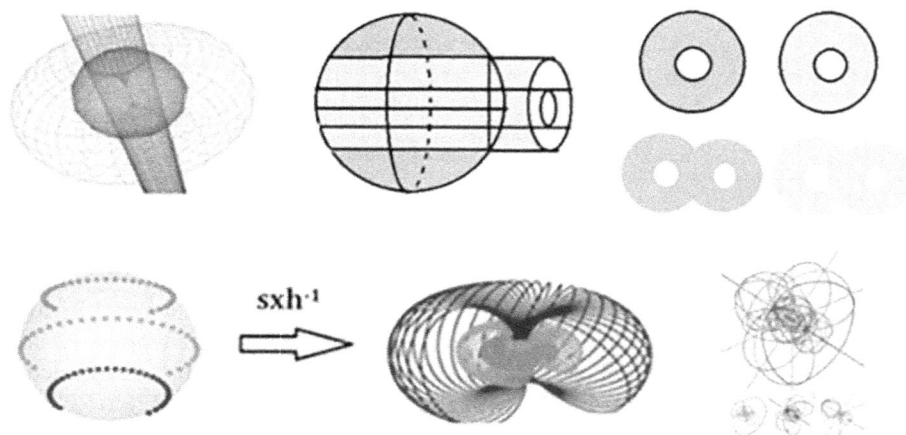

sxh⁻¹

Figure 9.
Translation from sphere to torus. Hopf fibration.

Villarceau circles." Every point on the sphere gives a torus and more points mean more tori, resulting in Hopf fibration characteristics for *cis-* and *trans*-stereochemistry in multidimensional space [18].

The main question, in case of six sets angles on two units U and S, from the transformation U to S results in a torus or inverse of torus (Dupin cyclide), since unit S is sometimes almost equal with *tan* function. Probably torus invers of torus is the most expected relationship.

Topologically, a ring torus is homeomorphic to the Cartesian product of two circles: $T^2 = S^1 x S^1$, a compact 2-manifold of genus 1, a modified version of the spherical coordinates system – having toroidal and poloidal coordinates, with cylindrical coordinates, a limit case of the conic section. Considering a unit 3D cube, once bringing the opposite faces together results in a torus with four characteristic circles: parallels, meridians, and a Villarceau circle, resulting in the intersection of the torus with a bitangent plane [5, 9].

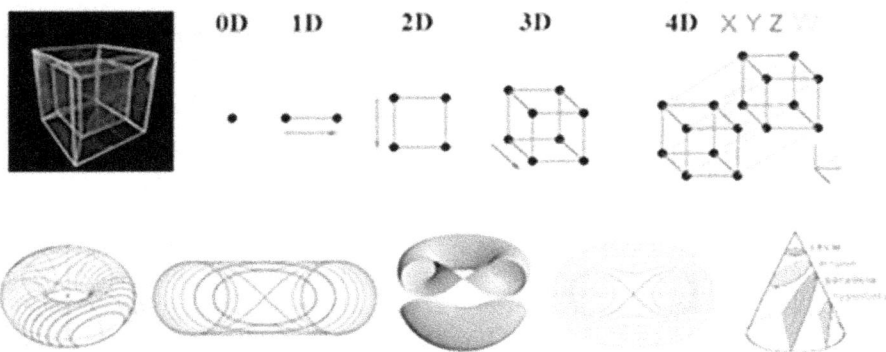

Figure 10.
Tesseract: 4D analog of the cube.

Cube and conic section two are basic equations used for the calculation of the first angle for building units. Toric section and conic section have representative polar curves at the intersection with a plane. Continuum transformation (**Figure 10**) from sphere to torus, then torus to rectangle is a process preserving homotopie, able to give many manifolds – point, line, circle, rectangle, sphere, torus. Lines of rectangle bend into circle, circle into cylinder, and cylinder into torus [5, 9].

2.8 Manifolds on 3-sphere approach

Manifolds used to date on 3-sphere approach: Euler-conic (18), conic section, rectangle, cyclide, and polar equations. Java Script programs enable calculation of one angle for building units from carbon or proton with Euler conic, conic section, from carbon and proton chemical shift in case of rectangle approach, and cyclide – Villarceau circles under *sin* and *tan* function of torus and Dupin cyclide, and from the ratio between differences in chemical shift between two atoms of carbon or two protons and vicinal coupling constant in case of polar equations [26, 27].

3-Sphere approach for the calculation of dihedral angles, a method in four steps:

1. calculation from recorded vicinal coupling constant $^3J_{HH}^{exp}$ [Hz] the vicinal angle ϕ^{exp} [deg] and corresponding dihedral angle θ_{HH}^{exp} [deg],

2. calculation from chemical shift δ [ppm] with manifold equations one angle θ^{CA} [deg] (C – U or S) of set A usefully for building units,

3. choosing dihedral angle θ_{HH} [deg] with value almost equal with dihedral angle $\theta_{HH}{}^{exp}$ [deg] calculated from recorded vicinal coupling constant $^3J_{HH}{}^{exp}$ [Hz],

4. calculation the vicinal coupling constant $^3J_{HH}$ [Hz] from vicinal angle ϕ [deg] calculated from dihedral angle θ_{HH} [deg].

2.9 Endocyclic and exocyclic dihedral angles

Endocyclic angles and exocyclic angles calculated with Karplus equations are under +/ −120 [deg], or endocyclic from exocyclic with PSEUROT equations or program, method which give the corresponding sign for every stereochemistry. 3-sphere dihedral angles are considered exocyclic angles with right sign and stereochemistry result from trigonometric rule characteristic of wave NMR data, which is in agreement with algebraic equations. On molecular models, 3-sphere endocyclic angles have different signs for *trans-ee* stereochemistry as simulate with VISION molecular models the phase angle of pseudorotation P [deg]. Gausian09W is working properly with endocyclic angles [28].

3. 3-Sphere tetrahedral angles of six- and five-membered ring

Platonic solids or Catalan solids, with pentagon faces, are homeomorphic to the sphere. Poincaré dodecahedron space is a binary icosahedral group of order 120. Dihedral angles of tetrahedrons are not commensurable with 2π, a hole with a common edge remains between two faces, and perfect tiling of the Euclidean space R^3 is not possible with regular tetrahedrons. Irregular icosahedrons with edge lengths slightly longer than circumsphere radius r (l = 1.05r) result from 20 tetrahedra, having as solution spherical space, not Euclidean. They look locally as 3D Euclidean models. Hundred and 20 vertices belong to hypersphere S^3 with golden ratio as radius. The shape of natural organisms, that is, viruses and carbasugar, are governed by Fibonacy number, Vedic nuclear Physics, and Tesla numbers. The shape of viral capsid is either helical or icosahedral, and five- and six-membered rings natural product also under dodecahedron and icosahedron geometries. The Fibonacci ratio 1.6190476 is closely approximating with Phi number 1.6180339, at list with dodecahedron 1.618029 and icosahedron 2.6180178 [4, 5].

dihedral angle θ_{HnHn+1}[deg]
vicinal angle ϕ[deg]

Fall ACS Boston 2018

tetrahedral angle φ_{Cn}[deg]
internal angle γ[deg]

Fall ACS Chicago 2022

Figure 11.
Polygon and three-sphere on calculation four angles: vicinal-dihedral and tetrahedral and internal of five- and six-membered rings.

Sclafi Hess polygon, icosahedron with 120 cell {3, 5, 5/2}, a polygon with a five-membered ring unit – pentagon (72, 144), and a six-membered ring unit – hexagon (60, 120), on 3-sphere ensuring greater relationships between dihedral angles $\theta_{HnHn + 1}$[deg] and tetrahedral angles φ_{Cn} [deg] of five- or six-membered rings, calculated from NMR data, only from vicinal coupling constant $^3J_{HH}$ [Hz] or chemical shift δ_{Cn} [ppm] (**Figure 11**) [25].

3.1 3-Sphere tetrahedron angles calculated from NMR data

Three possible dihedral angles can be calculated from only one vicinal coupling constant, and all are in close relationship with one pair of tetrahedral angles φ_{Cn}, φ_{Cn+1} [deg]. The relationship between dihedral-tetrahedral – phase angle of the pseudorotation decreased the number of dihedral angles and listed the number of possible conformations.

3.1.1 Relationships for calculation tetrahedral angles of a five-membered ring

Relationships between dihedral θ_{HH} [deg] and tetrahedral φ_{Cn} [deg] angles on seven sets angles, calculated from *sin* or *tan* values of manifold in π, can be established using angles θ^{A1}, θ^{A2} or θ^{B1}, θ^{B2} for calculation θ^{D4}, θ^{E4}, or θ^{F4}, θ^{G4}. Then, tetrahedral angles φ_{Cn} [deg] are chosen in opposite (*sin* versus *tan*) with dihedral angles θ_{HH} [deg]. Angles in opposite results also from transformation U to S and S to U [29, 30]:

$$2D : A = \sin^{-1}\left(\frac{1}{Rm}\right) \tag{62}$$

$$2D : B = \tan^{-1}\left(\frac{1}{Rm}\right) \tag{63}$$

$$3D : \varphi_{Cn} = \cos^{-1}\sin\left(\frac{-A \text{ or } -B}{2}\right) \tag{64}$$

where R_m – [π] – values calculated from chemical shift δ [ppm] with manifold equation, φ_{Cn} – tetrahedral angles with C_n = C_1, C_2, C_3, and C_4.

Relationships between dihedral angles of a five-membered ring and tetrahedral angles on seven sets angles particularized for *cis*[6,1 and 5,2] stereochemistry:

I. *Cis, trans-aa*[6,1] stereochemistry:

$$\cos^{-1}\sin\left(-\theta^{A1}/2\right) = \varphi_{C2,C3} \tag{65a}$$

$$\cos^{-1}\sin\left(-\theta^{A2}/2\right) = \varphi_{C1,C4} \tag{65b}$$

if θ^{A1} = θ_{HH} – dihedral angles with *cis-aa*[6,1] stereochemistry [deg].

II. *Cis, trans-aa*[5,2] stereochemistry:

$$\cos^{-1}\sin\left(\theta^{B2}/2\right) = \theta^{A2} \tag{66a}$$

$$\cos^{-1}\sin\left(-\theta^{A2}/2\right) = \varphi_{C1,C4} \tag{66b}$$

$$\cos^{-1}\sin\left[-\left(60-\theta^{A2}\right)/2\right] = \varphi_{C2,C3} \tag{66c}$$

if θ^{A2} = ϕ – vicinal angle ϕ [deg], θ^{B2} = θ_{HH} – dihedral angles with *cis-aa*[6,1] stereochemistry [deg].

III. *Trans-ee*[3,2 or 4,1] stereochemistry:

$$\cos^{-1}\sin\left[-\left(\theta^{A3}-60\right)/2\right] = \varphi_{C2,C3} \tag{67a}$$

$$\cos^{-1}\sin\left[-\left(\theta^{A4}-60\right)/2\right] = \varphi_{C1,C4} \tag{67b}$$

If $\theta^{A3, \, A4} = \theta_{HH}$ – dihedral angles with *trans-aa* stereochemistry [deg].

3.1.2 Relationships for calculation tetrahedral angles for six-membered rings

Tetrahedral angles of a six-membered ring are calculated with (Eqs. 68–70) and (71) from carbon and/or proton chemical shift. Seven sets of units are built from calculated tetrahedral angles from *sin* and *tan* function. Relationships between tetrahedral and dihedral angles are established between five units, compared with five-membered rings where the opposite rule was preferred studied some consecutive vicinal coupling constants. Because units starting with first angles of sets A and B smaller as 5[deg] or higher as 5[deg] are successions of unit S or U, and the relationship between U and S is almost under *sin* and *tan* function, in case of six-membered rings tetrahedral and dihedral angles are chose from seven sets unit [30, 31]:

$$\textbf{2D} : \sin^{-1}[nx(1/R_m)] = X^S \tag{68a}$$

$$\textbf{3D} : \sin^{-1}\left[mx\tan\left(-n_a x X^S\right)\right] = -\phi_2^F, m = 1, \tfrac{1}{2}, n_a, n_b = \tfrac{1}{2}, 1, 2 \tag{68b}$$

a. if $A = \theta^{C3}/2 < 59.9[\text{deg}]$, $A = \theta^{C2}/2 = (120 - \theta^{C3})/2$, $\cos^{-1}\sin\text{-}[(120 - \theta^{C3})]/2 = \cos^{-1}\sin\text{-}(\theta^{B1}) = \varphi_{Cn}^{A4}$;

where two pairs of sets angles D, E or F, G bearing dihedral and vicinal angles, algebraic angle $\phi_2^{F \text{ or } D}$

b. $\phi_2^B = \theta^{C1} < 29.9[\text{deg}]$, $\cos^{-1}\sin\text{-}(\theta^{E1})/2 = \varphi_{Cn}^{D4}$

where two sets angles A, B bearing dihedral and vicinal angles, algebraic angle ϕ_2^B

$$\textbf{2D} : \tan^{-1}[nx(1/R_m)] = X^{IT} \tag{69a}$$

$$\textbf{3D} : \tan^{-1}\left[mx\sin\left(-X^{IT}/n_a\right)\right] = -\phi_2^F, m = 1, \tfrac{1}{2}, n_a, n_b = \tfrac{1}{2}, 1, 2 \tag{69b}$$

c. if $\varphi_{Cn} = \varphi_{Cn}^{A5}$ and $\phi_2^G = \theta^{B2} < 29.9[\text{deg}]$, $\phi_2^F = \theta^{C}{}^2/2$, $\cos^{-1}\sin\text{-}(\phi_2^F) = \varphi_{Cn}^{A4}$

$$180 - \varphi_{Cn}^{A5} = \theta^{A2}, 60 + \theta^{A2} = \varphi_{Cn}^{A4}$$

$$\textbf{2D} : \tan^{-1}[nx(1/R_m)] = X^T \tag{70a}$$

$$\textbf{3D} : \tan^{-1}\left[mx\sin\left(-X^T/n_a\right)\right] = -\phi_2^F, m = 1, \tfrac{1}{2}, n_a, n_b = \tfrac{1}{2}, 1, 2 \tag{70b}$$

Eqs. (68a,b)–(70a,b) in 2D and 3D can be written in 4D (71a,b), hypersphere generalized equations for calculation angles required for Eq. (72):

$$\textbf{4D} : \sin^{-1}\tan\left\{-\sin^{-1}[nx(1/R_m)]\right\} = -\phi_2^{FS} \tag{71a}$$

$$\textbf{4D} : \tan^{-1}\sin\left\{-\tan^{-1}[nx(1/R_m)]\right\} = -\phi_2^{FT \text{ or } FIT} \tag{71b}$$

$$\textbf{3D} : \cos^{-1}\sin\left[-n_b x \phi_2^F\right] = \varphi_{Cn} \tag{72a}$$

where φ_{Cn} [deg] – tetrahedral angle of C_n, n = 1–6, R_m – [π] – values calculated from chemical shift δ [ppm] with manifold equation.

The best relationship between the tetrahedral angle (angle A4 on set A) and angle C2 on set C is Eq. (72b), able to establish the relationship between the first angle of set B and the tetrahedral angle, fourth angle of set A.

$$\textbf{3D}: \cos^{-1}\sin\left[-mx\theta^{UC2}\right] = \varphi_{Cn}, m = 1/2 \tag{72b}$$

Relationships between tetrahedral φ [deg] and vicinal ϕ [deg] angles of a six-membered ring on seven sets angles, relationships between sets A, B and F, G and A, B, and D, E:

$$\theta^{A1} = \delta_{Cn}\text{--}120, \theta^{A2} = \delta_{Cn}\text{--}60, \tag{73ab}$$

$$\cos^{-1}\sin\varphi_{Cn} = \theta^{B1}, \theta^{B2} = 60\text{--}\theta^{B1}, \tag{73cd}$$

where: $\varphi_{Cn} = \theta^{A4}$

I. *cis, trans-aa*[6,1] stereochemistry:

$$\sin^{-1}\cos\left[(60\text{--}\theta^{A1})/2\right] = \theta^{D3} = \phi^{6,1} \tag{74a}$$

$$\sin^{-1}\cos\left[(60\text{--}\theta^{A2})/2\right] = \theta^{E3} = \phi^{6,1} \tag{74b}$$

$$\sin^{-1}\cos\left[(60\text{--}\theta^{B1})/2\right] = \theta^{F3} = \phi^{6,1} \tag{74c}$$

$$\sin^{-1}\cos\left[(60\text{--}\theta^{B2})/2\right] = \theta^{G3} = \phi^{6,1} \tag{74d}$$

II. *cis, trans-aa*[5,2] stereochemistry:

$$\sin^{-1}\cos\left\{60\text{--}\left[(60\text{--}\theta^{A1})/2\right]\right\} = \theta^{D2} = \phi^{5,2} \tag{75a}$$

$$\sin^{-1}\cos\left\{60\text{--}\left[(60\text{--}\theta^{A2})/2\right]\right\} = \theta^{E2} = \phi^{5,2} \tag{75b}$$

$$\sin^{-1}\cos\left\{60\text{--}\left[(60\text{--}\theta^{B1})/2\right]\right\} = \theta^{F2} = \phi^{5,2} \tag{75c}$$

$$\sin^{-1}\cos\left\{60\text{--}\left[(60\text{--}\theta^{B2})/2\right]\right\} = \theta^{G2} = \phi^{5,2} \tag{75d}$$

III. *trans-ee* stereochemistry:

$$\sin^{-1}\cos\left[60 + (\theta^{A1})/2\right] = \theta^{E1} = \phi^{3,2 \text{ or } 4,1} \tag{76a}$$

$$\sin^{-1}\cos\left[60 + (\theta^{A2})/2\right] = \theta^{D1} = \phi^{3,2 \text{ or } 4,1} \tag{76b}$$

$$\sin^{-1}\cos\left[60 + (\theta^{B1})/2\right] = \theta^{G1} = \phi^{3,2 \text{ or } 4,1} \tag{76c}$$

$$\sin^{-1}\cos\left(60 + (\theta^{B2})/2\right) = \theta^{F1} = \phi^{3,2 \text{ or } 4,1} \tag{76d}$$

All the calculated vicinal angles ϕ[deg] are for values < or = 180[deg], 180[deg] will be added if the calculated ϕ from $^3J_{HH}$[Hz] (eq. 48, 49) is higher as 180[deg].

3.2 Carbon chemical shift calculated from tetrahedral angles

Carbon chemical shift of the five-membered ring results from the relationship between ϕ_2 of sets D, E, the first angle of set A, and ϕ_2 of sets F, G, the first angle of set

B. Building the other three sets of units from I. D, E, and II. F, G, angles of set A and B are corresponding angles of set C for units I and II:

$$\phi_{1/2}{}^{D} = \cos^{-1} \sin \varphi_{Cn}, \text{ with } C_n = C_2, C_3 \tag{77}$$

$$\phi_{1/2}{}^{E} = \cos^{-1} \sin \varphi_{Cn}, \text{ with } C_n = C_1, C_2 \tag{78}$$

where $2x\phi_{1/2}{}^{D} = \phi_{2}{}^{E} = \theta^{A2}$
$2x\phi_{1/2}{}^{E} = \phi_{2}{}^{D} = \theta^{A1}$.
$\phi_{2}{}^{D}, \phi_{2}{}^{E}$ angles θ^{A1}, θ^{A2} of set A

$$\sin^{-1}\theta^{A1} \text{ or } \theta^{A2} = 1/Rm, \tag{79}$$

$$\tan^{-1}\theta^{A1} \text{ or } \theta^{A2} = 1/Rm \tag{80}$$

where R_m can be calculated from θ^{A1} or θ^{A2} in the function of the vicinal coupling constant.

Carbon chemical shift of a six-membered ring can be calculated from tetrahedral angles using Eq. (72b). Euler-conic as manifold (18) transform R_m from radieni (gauss) in ppm:

$$\mathbf{3D}: \tan^{-1}\left[\mathrm{mxsin}\left(-\mathrm{nx}\theta^{UC2}\right)\right] = \theta^{S} \tag{81a}$$

$$\mathbf{3D}: \sin^{-1}\left[\mathrm{mxtan}\left(-\mathrm{nx}\theta^{UC2}\right)\right] = \theta^{Ta} \tag{81b}$$

$$\mathbf{3D}: \sin^{-1}\left[\mathrm{mxtan}\left(-\mathrm{nx}\theta^{UC2}\right)\right] = \theta^{Tb} \tag{81c}$$

$$\mathbf{3D}: \sin^{-1}\theta^{S} = Rm \tag{82a}$$

$$\mathbf{3D}: \tan^{-1}\theta^{Ta} = 1/Rm \tag{82b}$$

$$\mathbf{3D}: \tan^{-1}\theta^{Tb} = Rm \tag{82c}$$

where R_m can be calculated from θ^{S}, θ^{Ta}, or θ^{Tb} in function of vicinal coupling constant.

3.3 3-Sphere approach on conformational analysis

Conformational analysis, the phase angle of the pseudorotation P [deg] and the angle of deviation from planarity θ_m [deg] are calculated for five [32] and six [25] membered rings with the Altona model and 3-Sphere dihedral angles, exocyclic angles instead of endocyclic angles.

On VISION molecular models, the phase angle of pseudorotation P [deg] can be visualized using endocyclic angles, three-sphere dihedral angles in this case. From the three possible dihedral angles for every vicinal coupling constant, based on the stereochemistry of carbasugar, signs of dihedral angles are excluded. In a strict sense, the three-sphere approach is not a method for the calculation of only dihedral angles, and it is a method for the calculation of a pair of angles, dihedral-vicinal and tetrahedral and internal, having a close relationship with phase angle of pseudorotation [28].

Other application of three-sphere dihedral angles: Conformation and configuration at anomeric position [33] can be analyzed with a ratio of anomers and three-sphere-Lambert-Wu methods for calculation of dihedral angles from NMR data. From three-sphere dihedral angles can be calculated the lengths between two atoms of carbon or distances between two protons [34].

4. Conclusions

3-Sphere, a hypersphere in higher dimension, ensures calculation of dihedral θ_{HnHn+1} [deg] and tetrahedral φ_{Cn} [deg] angles with trigonometric and algebraic equations under Hopf fibration and Lie algebra theories. Relationships between Lie Algebra E8 and Hopf fibration with key sphere H7. Units are built from angles calculated θ^{CA} [deg] only from the vicinal coupling constant or from manifolds. On seven sets angles are found two pairs of angles: vicinal ϕ [deg] – dihedral θ_{HH} [deg] and tetrahedral φ_{Cn} [deg] and internal γ [deg].

From only one vicinal coupling constant $^3J_{HH}$ [deg] results in three possible dihedral angles θ_{HH} [deg] with right sign and stereochemistry along to corresponding characteristics tetrahedral angles φ_{Cn} [deg], apparently too many but can be excluded based on stereochemistry of sugar and Newman representation on VISION molecular models.

Carbon chemical shift can be calculated for five- and six-membered rings from relationships for calculation tetrahedral angles. In case of a six-membered ring directly from a tetrahedral angle and in case of a five-membered ring from first or second angles of sets A or B in the function of vicinal coupling constant.

Conflict of interest

The authors declare no conflict of interest.

Acronyms and abbreviations

NMR   nuclear magnetic resonance

Author details

Carmen-Irena Mitan[1]*, Emerich Bartha[1], Petru Filip[1] and Robert Moriarty[2]

1 Institute of Organic and Supramolecular Chemistry "C. D. Nenitescu", IL, Bucharest, Romania

2 Organic Chemistry, University of Illinois at Chicago, Chicago, IL, USA

*Address all correspondence to: cmitan@yahoo.com

IntechOpen

References

[1] Bartha E, Mitan C-I, Draghici C, Caproiu MT, Filip P, Moriarty R. Program for prediction dihedral angle from vicinal coupling constant with 3-sphere approach. Revue Roumaine de Chimie. 2021;**66**:178-183. DOI: 10.33224/rrch.2021.66.2.08

[2] Mitan C-I, Bartha E, Filip P, Draghici C, Caproiu MT, Moriarty RM. NMR data on conformational analysis of five and six membered ring under 3-sphere approach. Vicinal constant coupling $^3J_{HH}$ on relationships between dihedral angles and tetrahedral angles. In: Sustainability in a Changing World. ACS National Meeting in Chicago, IL, August 21-25, 2022, CARB 3717557. Washington, D. C: American Chemical Society; 2022. 37 p. DOI: 10.1021/scimeetings.2c00876

[3] Mitan C-I, Bartha E, Filip P, Draghici C, Caproiu MT, Moriarty RM. Two isomers with trans-aa^{5,2} stereochemistry are calculated with 3-sphere trigonometric equations approach at circle inversion motion from NMR data. In: Sustainability in a Changing World ACS National Meeting in Chicago, IL, August 21-25, 2022, CARB 3717658, 2022, Sci-Mix. 2022. DOI: 10.1021/scimmetings.2c00523

[4] Mitan C-I, Bartha E, Filip P, Caproiu M-T, Draghici C, Moriarty RM. Graph flux intensity and electromagnetic wave on 3-sphere approach science. Journal of Chemistry. 2023;**11**:212. DOI: 10.11648/j.sjs.20231106.12

[5] Available from: en.wikipedia.org/wiki/Keywords

[6] O'Sullivan B. The Hopf fibration and Hidden variables in quantum and classical mechanics. arXiv: 1601.02569v10[physics.gen-ph]. 2021

[7] O'Sullivan B. Quaterions, Spinors and the Hopf Fibration: Hidden variables in classical Mechanics, arXiv:1601.02569v14 [physics: gen-ph]. 2021

[8] Waite JV. The Hopf Fibration and Encoding Torus Knots in Light Fields. Las Vegas: University of Nevada; 2016. DOI: 10.34917/9112204

[9] Mitan C-I, Moriarty RM, Filip P, Bartha E, Caproiu MT, Draghici C. Conformational analysis on five membered ring by nuclear magnetic resonance spectroscopy. Relationships between constant couplings $^3J_{HH}$, chemical shifts and dihedral angles. In: 256th ACS National Meeting in Boston, MA, August 19–23, 2018. CARB 84, ID: 2972261. 51 p. Washington, D. C: American Chemical Society;

[10] Sweeney JF. Hopf Fibration and E8 in Vedic Physics, Vixra:1405–0358[Pdf]. High Energy Particle Physics; 2014. 34 p

[11] Kirillov A. Introduction to Lie Groups and Lie Algebras. Available from: http://www.math.sunysb.edu/"kirillov

[12] Mitan C-I, Bartha E, Draghici C, Caproiu MT, Filip P, Moriarty RM. Hopf fibration on relationship between dihedral angle θ_{HnHn+1}[deg] and vicinal angle ϕ[deg], angles calculated from NMR data with 3-sphere approach and Java script. Science Journal of Chemistry. 2022;**10**:21. DOI: 10.11648/j.sjc.2022 1001.13

[13] Mosseri R, Dandoloff R. Geometry of entangled states, Bloch spheres and Hopf fibration. Journal of Physics A: Mathematical and Theoretical. 2001;**34** (47):10243-10252; arXiv: quent-ph/0108137v1, 30 Aug. 2001.

[14] Ruf B, Srikanth PN. Concentration on Hopf fibration for singularly

perturbed elliptic equations. Journal of Functional Analysis. 2014;**267**:2353. DOI: 10.1016/j.jfa.2014.07.018

[15] Grigorian S. Algebraic structures on parallelizable manifolds. Journal of Algebra. 2024;**657**:804. DOI: 10.1016/j. jalgebra.2024.06.001

[16] Klim J, Majid S. Hopf quasigroup and the algebraic 7 sphere. arXiV: 0906.5026v3(math.QA). 2009

[17] Ornea L, Parton M, Piccinni P, Vulturescu V. arXiV: 1208.0899v2 [math.DG]. 2012

[18] Lyons DW. An elementary introduction to the Hopf fibration. Mathematic Magazine. 2003;**76**:87. DOI: 10.230713219300. Available from: lyons@lvc.edu

[19] Saito S. Active SU(2) operation on Poincare sphere. Results in Physics. 2024;**59**:107567. DOI: 10.1016/j. rinp.2024.107567

[20] Mitan C-I, Bartha E, Petru F, Draghici C, Caproiu M-T, Moriarty RM. Manifold inversion on prediction dihedral angle from vicinal coupling constant with 3-sphere approach. Revue Roumaine de Chimie. 2023;**68**:3, 185-4. DOI: 10.33224/rrch.2023.68.3-4.08

[21] Nilov F, Skopenkov M. A surface containing a line and a circle through each point is a quadric, arXiV: 1110.2338v2 [math.AG]. 2012

[22] Moriarty RM, Mitan C-I, Gu B, Block T. Hypersphere and antiviral activity of three alkyl chain innocyclitols with D and L Ribitol stereochemistry. American Journal of Heterocyclic Chemistry. 2023; **9**(1):9-24. DOI: 10.116481/ jajhc20230901.12. ISSN:2575-7059 (Print), ISSN:2575-5722 (online)

[23] Mitan C-I, Bartha E, Filip P, Draghici C, Caproiu MT, Moriarty RM. 3-sphere dihedral angles under wave character of the NMR data with applications on conformational analysis. In: ACS National Meeting in San Diego, CA, March 23–27, 2025. CARB: American Chemical Society, Washington, D. C. Poster Id 4175487

[24] Humphreys J. Introduction to Lie Algebras and Representation Theory1972. p. 42. ISBN: 0387900535

[25] Bartha E, Mitan C-I, Filip P. 3-sphere torsional angles and six membered ring conformation. American Journal of Quantum Chemistry and Molecular Spectroscopy. 2023;**7**(1):9. DOI: 10.116481j.ajqcms.20230701.12

[26] Bartha E, Mitan C-I, Draghici C, Caproiu MT, Filip P, Moriarty RM. Rectangle as manifold on relationships between vicinal constant couplings $^3J_{HH}$, ^1H and ^{13}C-chemical shifts and dihedral angles. Revue Roumaine de Chimie. 2022;**67**:167-172. DOI: 10.33224/ rrch.2022.67.3.05

[27] Mitan C-I, Bartha E, Filip P, Draghici C, Caproiu M-T, Moriarty RM. Java script programs for calculation of dihedral angles with manifold equations. Science Journal of Chemistry. 2024;**12** (3):42. DOI: 10.11648/j.sjc.20141203.11

[28] Filip P, Mitan C-I, Bartha E. 3-sphere tetrahedral angles and phase angle of the pseudorotation P[deg] of C_1-CH_3-α-D ribitol iminocyclitol. Science Journal of Chemistry. 2024;**12**:54. DOI: 10.11648/j. sjc.20241203.12

[29] Mitan C-I, Bartha E, Filip P. Relationship between tetrahedral and dihedral on hypersphere coordinates. Revue Roumaine de Chimie. 2023;**68** (5–6):261. DOI: 10.33224/rrch.2023.68.5-6.09

[30] Moriarty RM, Mitan C-I, Bartha E, Filip P, Naithani R, Block T. 3-Sphere approach on 9-O-(10,11-di-O-benzyl-12,14-O-benzylidene-α-D-galactopyranosyl)-1-butyl-2,3-O-isopropylidene-1,4-dideoxy-1,4-imino-1-N-dehydro-L-ribitol. American Journal of Quantum Chemistry and Molecular Spectroscopy. 2024;**8**(1):1. DOI: 10.11648/ajqcms.20240801.11

[31] Mitan C-I, Bartha E, Filip P. Tetrahedral angles of six membered ring calculated from NMR data with 3-sphere approach. Revue Roumaine de Chimie. 2023;**68**(5–6):273. DOI: 10.33224/rrch.2023.68.5-6.10

[32] Mitan C-I, Bartha E, Filip P, Draghici C, Caproiu MT, Moriarty RM. Dihedral angles calculated with 3-sphere approach as integer in conformational analysis on D-, L- ribitol series. Revue Roumaine de Chimie. 2022;**66**(21):941. DOI: 10.33224/rrch.2021.66.12.07

[33] Mitan C-I, Bartha E, Hîrtopeanu A, Stavarache C, Draghici C, Caproiu MT, et al. Configurational and conformational analysis of 5-deoxy-5-iodo-α,β-D-ribose with 3-Sphere approach. American Journal of Quantum Chemistry and Molecular Spectroscopy. 2023;**7**(1):1-8. DOI: 10.11648/j.ajqcms.20230701.11

[34] Mitan C-I, Bartha E, Filip P. Distances $L_{HnHn+1}[A^0]$ calculated from 3-sphere dihedral angles. Revue Roumaine de Chimie. 2024;**69**:579. DOI: 10.33224/rrch.2024.69.10-12.07

Chapter 3

Operando NMR of Adsorbed Molecules Becomes Feasible Using Benchtop Spectrometers

Deniz Uner and Ecem (Volkan) Deveci

Abstract

NMR of adsorbed molecules provides critical information in terms of catalytic reactions and their bottlenecks. This chapter will provide evidence that liquid NMR spectrometers can bear the capacity to measure adsorbed molecules and suspended solids. Using a 400 MHz liquid NMR spectrometer, $^{13}CH_4$ adsorption could be observed over a Ni/CeO_2 catalyst. A valve-sealed sample tube was used to pretreat the catalyst and seal it under an exposure of $^{13}CH_4$. Using the same spectrometer, it was possible to detect ^1H signals from graphene oxide particles suspended in D_2O, reproducing the results of earlier publications. The signals appeared to correspond precisely to the published data obtained from ^1H magic angle spinning (MAS) solid-state NMR spectroscopy. These demonstrated potentials encouraged us to pioneer coupling a low-field (1T) benchtop spectrometer to a gas manifold, to study adsorbed molecules over catalysts through ^1H and ^{13}C NMR spectroscopy. Our research was supported through collaborations using DFT NMR: unique proton environments were revealed in hydrides of Pd. New avenues of catalysis research are possible with benchtop NMR spectrometers, closely coupled to the gas manifolds, under vacuum and at high pressures.

Keywords: NMR spectroscopy of adsorbed molecules, supported metal catalysts, surface migration of adsorbed molecules, hydrogen spillover, benchtop spectrometers

1. Introduction

NMR of adsorbed molecules can offer significant insight into the catalytic processes [1]. NMR techniques help one to gather the chemical identity, the strength between the adsorbed molecule and the surface as well as the exchange rates of the nuclei among different environments. The ability to obtain such information in one single technique makes the method highly attractive for solving research problems relevant to surface reaction mechanisms. The primary difficulty of studying adsorbed molecules on solid catalysts stems from the limitations of high-resolution solid-state NMR spectroscopy. Signals from solid samples are subject to significant line broadenings due to factors such as chemical shift anisotropy, homonuclear and heteronuclear dipolar couplings, and couplings of the nuclei under NMR investigation to quadrupolar nuclei present in the sample.

The broadenings due to the effects mentioned above are not present in liquid systems. Molecules in the liquids undergo continuous and rapid tumbling, enabling the nuclei to experience an averaged-out and spherically symmetric field. Nuclei in the solid state, however, are in the same position relative to both static and field created by the pulses fluctuating with the radio frequency (RF field). As such, the final signal will have a distribution of frequencies due to nuclei with different relative orientations with respect to the static and fluctuating fields. This not only gives rise to broad signals that are difficult to detect, but their excitation requires stronger RF pulses.

Magic angle spinning is a strategy to overcome such line broadenings. The MAS technique introduced to the field by Andrew [2–4] has a well-understood physics: The Hamiltonian for chemical shift anisotropy and dipolar couplings as well as the central transitions of the quadrupolar coupling have the $(1-3 \cos^2\theta)$ term, where θ represent the angle between the magnetic field and the magnetic moment of the nuclei. Rotation of the sample can remove the broadening, only if the rotation is at such an angle θ, that it will render the $(1-3 \cos^2\theta)$ term zero, and at a frequency faster than the line broadening. The angle ensuring to eliminate $(1-3 \cos^2\theta)$ term is 54.74°, and the process is called magic angle spinning (MAS). The broad signals may require spinning rates of the order of several tens of kilohertz (kHz), and such rates are only possible with samples packed in very small diameter rotors.

Given this background of the NMR measurements in solids, one can envision the problems associated with the sample size and the inability to contact the sample with the reactive medium while the sample is spinning at very high rates in the magnetic field. The general principles of sample preparation strategies are summarized in review articles [5–7]. Flame-sealed samples in an NMR tube were widely used to monitor NMR spectra of adsorbed molecules, both under static conditions [8–10] and under MAS [11, 12].

Introduction of a gas flow to a sample that also undergoes magic angle spinning is a technical challenge [13–15]. This has been attempted to study reactions and adsorbed molecules. However, the problems associated with the samples spinning at very high rates while contacting a gas phase require careful management of gas-solid contact. Powders in a centrifugal force field will have a tendency to stick to the walls of the NMR rotor, and packing density can be much higher than the static counterpart. In the absence of the porosity due to the dense packing, the centrifugal force field can inhibit the flow of the gases that are introduced through a tube at the center of the rotor. In other words, the design has to be carefully screened for the disguises that limit the reaction monitoring in the kinetically relevant time scales (for a more detailed discussion, see, e.g., [15]).

While these challenges are present in high-resolution solid-state NMR of adsorbed molecules, one can explore if the surface mobility of the adsorbed molecules could be used to their advantage. Adsorbed molecules can undergo surface mobility (in 2D) and exchange between the surface and the gas phase (in 3D), fast enough to provide some motional averaging [11]. Early studies on sealed samples of ethylene over Ru/SiO_2 revealed that for highly mobile surface species, the solid-state techniques such as strong proton decoupling and magic angle spinning did not give rise to improved resolution [11, 12]. Instead, on these samples, application of the methods used in NMR of liquids such as observations of J splitting has proven to be useful in identifying the different types of carbon deposited on the surface. The experimental methodology using *in situ* NMR spectroscopy provided insight into several catalysis problems [1, 16]. The difference between solid and liquid-state NMR spectroscopies has created two systems that share the basics but differ in the applications. In the early days of the method, the same spectrometer could be used for both studying solids and liquids. As the technique has evolved with the introduction of the high-field superconducting magnets, the questions that can be answered with molecules in the liquid state under random motion and the questions that can be asked for the nuclei frozen in a solid state environment diverged, as well as their theories and the electronics. In time, these communities continued to address different needs in terms of resolution and working principles.

Before going deeper into the *in situ* and operando spectroscopy, some operational details about the pulses and data acquisition are in order. The RF power applied to a sample is equal to the product of the voltage (amplitude) of the RF and the current (duration) of the pulse in the time domain. The pulse NMR spectroscopy enabled the excitation over the full range of frequencies [17, 18] that is detected after the Fourier transform. While soft pulses (i.e., low voltage) of long durations are needed to obtain high resolutions in the liquid environments undergoing fast molecular tumbling, high-power sharp pulses were needed to excite the solid spectra with broad linewidths in the frequency domain.

Motional averaging has been shown through numerous studies to decrease the linewidths, even with small solid particles suspended in liquids [19]. The material covered so far, hopefully, convinces the reader that if the adsorbed molecules possess sufficient mobility, liquid NMR tools and strategies can enable a reasonable spectral resolution. Adsorbed molecules are unique in terms of their behavior, remaining between a solid and a liquid, depending on the bond strength and the coverage. The interaction between a surface and an adsorbed molecule can be solid-like or liquid-like, depending on the chemical environment [20]. The determining factor for this is the thermodynamics between the surface and the fluid phase. The exchange rate of the adsorbed molecules depends on both the ligand and steric factors on the surface [21].

The enthalpic effects reflect themselves in the characteristics of the bonds and events leading to the bond-making event. These factors can be easily elucidated through the chemical shifts, which can reveal the change in the strength of the bonds. The molecular orientation of the molecules and their mobility are embedded in the entropic terms. These influences are as important during chemical conversions as the enthalpic effects. As far as the entropic effects are concerned, elucidation is very difficult.

The entropic effects arise due to motion, a catalytic environment, and molecular orientations. While these facts are well acknowledged in catalysis, true information is accessible either through theory or through indirect measurements of adsorption isotherms and calorimetry. It is possible to reveal such entropic effects of the adsorbate molecules through NMR spectroscopy, through techniques that are widely used in molecular structure identification in organic chemistry. The bottlenecks in such studies are many, but we will begin with the broad lines without any structure, and the difficulties to prepare samples with a sufficient amount of adsorbate molecules under hermetically sealed conditions.

2. Infrastructure needed for *in situ* and *operando* NMR spectroscopy of adsorbed molecules

The high-field NMR spectrometer technology requires the sustenance of the superconductivity of the magnet coil in a cascade of cryogenic Dewars. The magnetic field created through the superconducting coil extends beyond the magnet boundaries. Therefore, any flow system involving components that can interact with the magnetic field has to be maintained at a sufficiently far distance. Furthermore, thermal management within the probe is challenging due to the required currents to induce resistive heating. As the need to access information that can only be accessed through NMR spectroscopy continues, various alternative solutions are being offered [5, 13, 15]. The issues of gas transfer are critical. For measurements under vacuum, the gas transfer line diameter becomes critical, as the mean free path of the molecules increases with decreasing gas pressure. Under the conditions that the mean free path is longer than the gas transfer line diameter, the molecular transport falls into a Knudsen diffusion regime, altering the quality of the *in situ* and operando measurements.

One of the early accounts of the *in situ* NMR measurements revealed a dynamic interaction of adsorbed hydrogen molecules over Ru/SiO_2 [16, 20]. Their home-built NMR probe was capable of handling temperatures up to 473 K, and it was connected to a vacuum manifold capable of high vacuums up to 10^{-6} Torr. The advantage of this type of close-coupled system was immediately realized: the effects of molecular mobility within the system could be monitored much better than the sealed samples. The limited amount of gas that can be sealed with the sample would have equilibrated by the time of the measurement. However, in the presence of a gas phase circulating throughout the system, it could reveal a significant level of information that was not accessible before.

While this system was highly sophisticated, a simpler adsorption strategy was possible with the NMR tubes with sealing options. Such tubes could easily be fitted to a gas manifold, the pretreatment and gas dosing of the samples could be carried out under appropriate temperature and pressure conditions, then the adsorbing gas is dosed, and the sample tube could be sealed for measurement in an NMR spectrometer, that is originally designed for liquid phase measurements. While measuring solid samples in liquid spectrometers sounds unusual, the fact is that the signal from surface species is generally heterogeneously broadened. In other words, the broad spectrum is a superposition of numerous sharp signals emanating from highly mobile surface species. This hypothesis is tested in a 400 MHz spectrometer for hydrogen adsorbed over Ru/SiO_2 and benzene adsorbed on silica gel.

Hydrogen adsorption over Ru/SiO_2, which is well studied under *in situ* conditions [8], was chosen to verify the suitability of the liquid spectrometers. The NMR spectrum characteristics of hydrogen adsorbed over Ru could be replicated using a valve-sealed sample. **Figure 1** shows NMR signal pertaining to OH groups from silica surface at resonances close to that of water ~5 ppm and shielded NMR signal from hydrogen adsorbed over Ru ~ −20–60 ppm. Due to the narrow spectral range of the measurement limited by the selected experimental conditions, it was not possible to record the signal emanating from H/Ru. But the broad asymmetric signal peaking at −25 ppm was clearly representative of H/Ru, responding to experimental conditions clearly designed for measurements in the liquid state.

The measurements for benzene adsorption over silica gel needed an internal standard for sample shimming. In order to achieve this, a capillary was filled with D_2O, and the residual ^1H signal in D_2O was used as the chemical shift calibration. The results are presented in **Figure 2**. As received, silica gel gave rise to a strong and broad signal, peaking at 7 ppm. This signal was lost after an overnight evacuation. When the sample was exposed to benzene vapor, ^1H NMR revealed a broad signal in close proximity where protons in benzene would resonate.

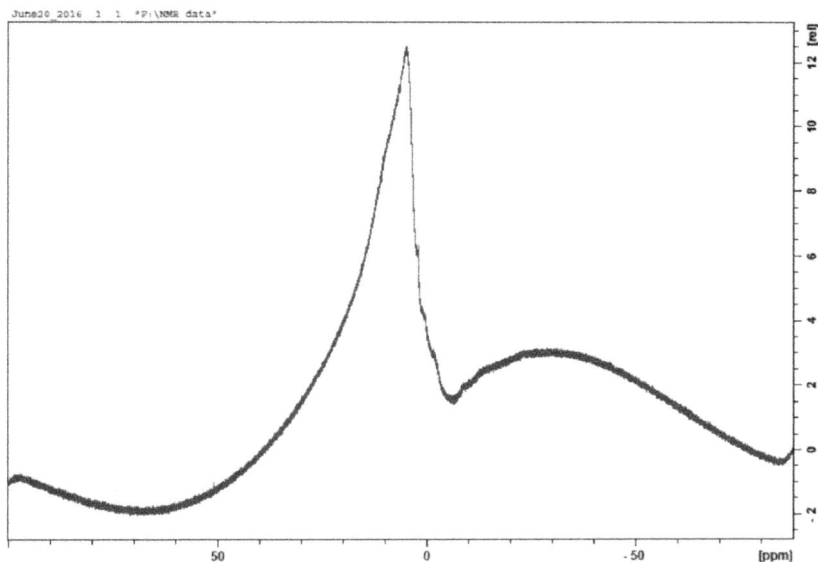

Figure 1.
The ¹H NMR measurement was performed over a valve-sealed NMR tube with 500 torr H_2. The features are not perfect since the measurement is performed using a liquid NMR spectrometer. The signal lineshape is perturbed by the cutoff of the time domain signal without full relaxation. Other than this, the signal is well correlated with the published NMR data. The measurement was taken using an AV400 Bruker spectrometer using a liquid probe.

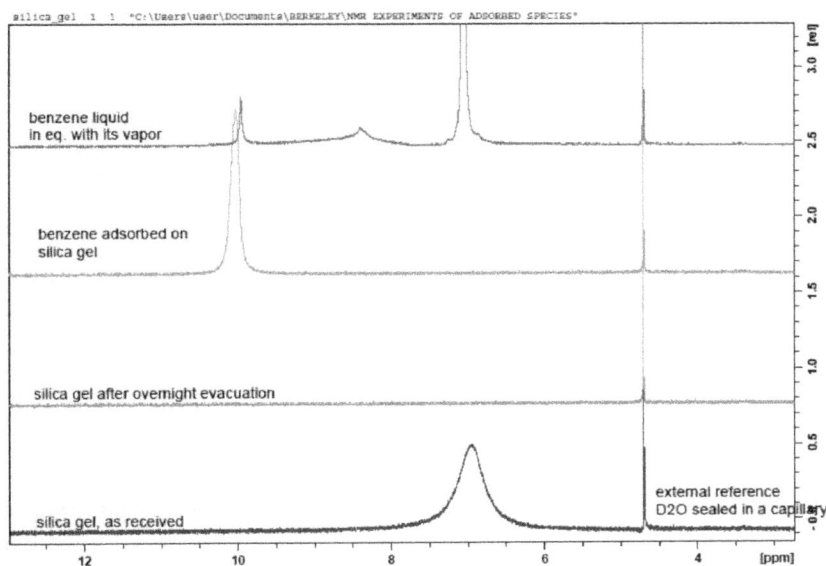

Figure 2.
Benzene adsorbed on silica gel was monitored through a sealed sample in a liquid NMR spectrometer. In order to shim and correctly reference the samples, D_2O with trace H_2O was sealed in a capillary and placed in the sample tube. The sharp signal at xxx ppm is due to the ¹H signal from H_2O in D_2O. The comparison of the benzene vapor in equilibrium with its liquid has three features. The leftmost feature at 10, x ppm is due to liquid benzene. The unusual shift is due to the electrons shared by benzene molecules in close-packed environments. The feature at 7.x ppm is due to the benzene vapor. This feature is very similar to the benzene fingerprint in liquid solvents at 7.15 ppm. The signal in the middle is due to the interface where liquid benzene and its vapor coexist. The measurement was taken using an AV400 Bruker spectrometer and a liquid probe.

The more interesting result was when an NMR tube introduced to a small amount of benzene liquid to monitor benzene in the vapor phase. The signals shown in the figure were surprising. The ^{1}H chemical shift of the protons in benzene is reported to be at 7.3 ppm. The sharp signal at 7.3 ppm was due to benzene in the liquid phase. The signal around 10 ppm corresponds to benzene molecules adsorbed on silica gel. A similar signal appears for benzene adsorbed on the walls of the NMR tube. The intermittent broad signal is tentatively assigned to benzene vapor.

3. NMR of adsorbed gases on supported metal catalysts

First, attempts to study adsorbed molecules with a liquid NMR spectrometer were performed on a 400 MHz spectrometer. Gas-phase methane resulted in a doublet. The reason was attributed to the perturbations arising due to the collisions with the tube wall. For the gas phase with low density, perturbations can be expected to cause a shifted signal that is different from the signal emanating from the gas-to-gas colliding species. The same measurement, performed this time with a fresh catalyst, gave rise to a significantly broadened spectrum for several reasons. First and foremost, methane has a high affinity toward metal surfaces, and adsorption will decrease the 3D mobility by one degree of freedom. On the surface, the molecule can only have 2D mobility and an occasional adsorption/desorption type of exchange. Magnetic susceptibility broadening is inevitable while studying adsorbed molecules. Interestingly, the catalyst sample with deposited coke did not give rise to any signal, indicating that adsorption over the coke-deposited catalyst was not possible (**Figure 3**).

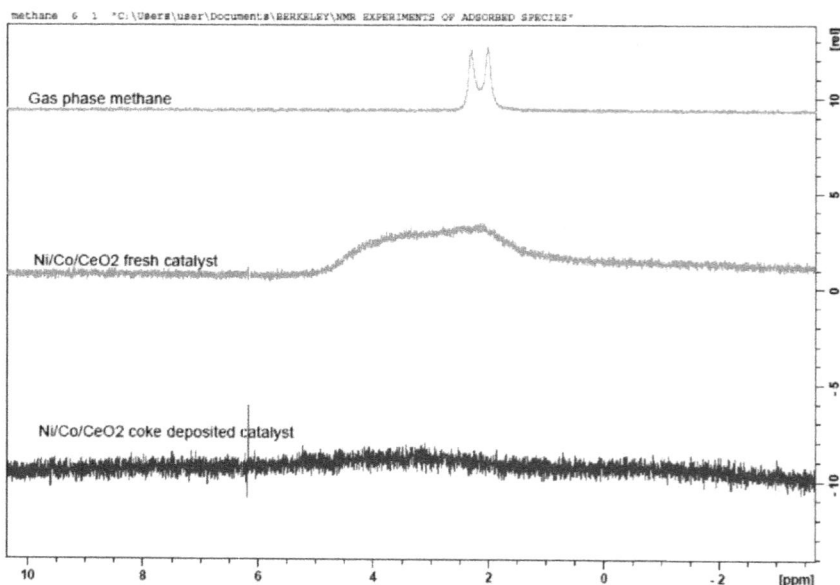

Figure 3.
^{1}H NMR data of ^{13}CH$_4$ over Ni-Co/CeO$_2$ catalyst. The gas phase peak splits due to the heteronuclear coupling between ^{1}H and ^{13}C nuclei. The peak shifts to the deshielded domain and broadens in the presence of a catalyst because of the magnetic susceptibility issues as well as the change in the behavior due to adsorption. Using a catalyst with coke deposition does not reveal any appreciable signal, indicating the absence of any adsorbate interactions accessible by this technique. The measurement was taken using an AVB400 Bruker spectrometer and a liquid probe.

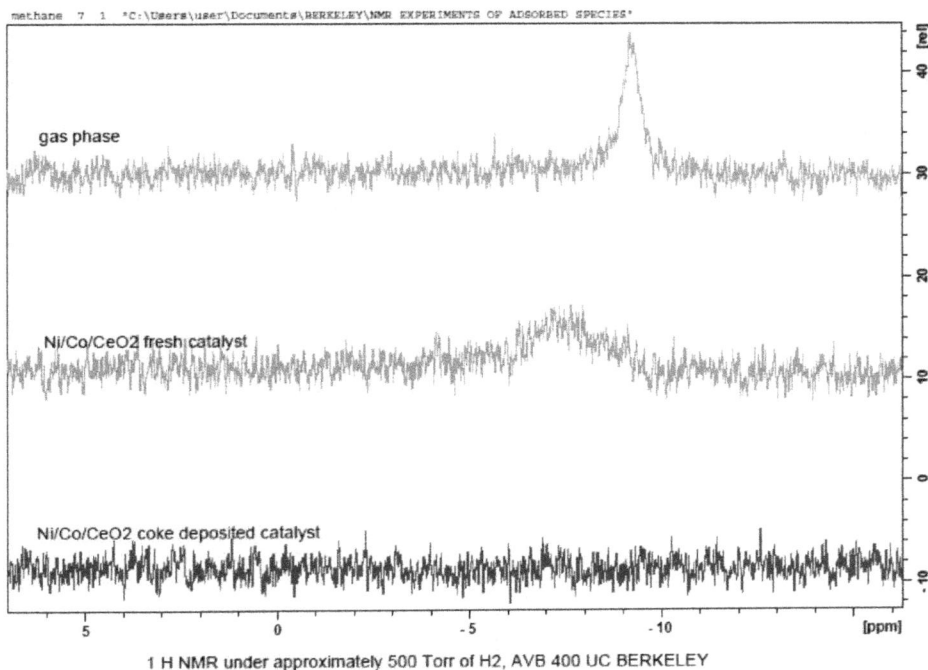

Figure 4.
^{13}C NMR spectra of methane in the gas phase and adsorbed on the catalyst. Similar to the ^{1}H signal, the ^{13}C signal is also absent on the coke-deposited sample. The measurement was taken using an AVB400 Bruker spectrometer using a liquid probe.

The doublet on the gas phase methane was attributed to the heteronuclear splitting of the ^{1}H signal due to the spin of ^{13}C [22, 23]. Similarly, ^{13}C measurements revealed a shift toward de-shielded region in the presence of a fresh catalyst. Similar to the ^{1}H signal, ^{13}C signal was also absent, confirming the absence of adsorption on the catalyst surface (**Figure 4**).

4. NMR of solids suspended in liquid phase

One final attempt to use a liquid spectrometer to characterize a solid was to monitor the signal from graphite. The graphitic parent material was sonicated in a deuterated solvent, and the supernatant fluid was taken for analysis. The expectation was to be able to have enough mobility in the suspended particles to observe a signal. The measurements revealed two signals that were consistent with their counterparts in the literature.

Pines group [19] has published an article on using the adsorbates on nanoparticles, assisting the Brownian motion of the particles in the liquid phase and allowing a liquid spectrometer to measure suspended solid particles. They reported the limits of such measurements: it, it is important to stay in the nanoscale in order for the adsorbed ligands to interact with the solvent molecules and induce molecular motion of the suspended particles.

The idea for the experiment with high surface area graphite suspended in D_2O was intended to demonstrate that liquid NMR techniques could offer insights into the

nanoparticles. In order to prepare these samples, solid graphene particles were sonicated for about 1 hour. After sonication, a waiting period was needed to let the heavy
particles settle. The supernatant solution, which is transparent but gray in color, was
taken to NMR for measurement. The results are shown in **Figure 5**. The data range
was selected to eliminate the signal coming from the residual water exchanged with
D_2O during the process. When the measured spectrum was compared to a highly
cited paper about NMR on graphene oxide, striking similarities were found when

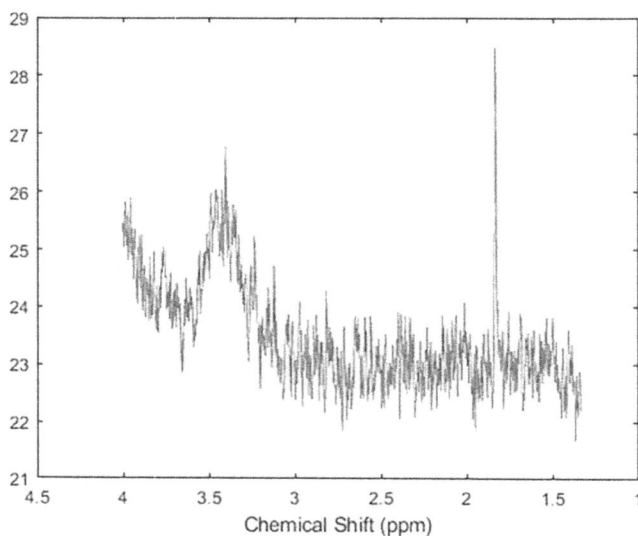

Figure 5.
*High surface area graphite was sonicated and centrifuged to produce suspended domains in D_2O. The
measurement was taken using an AVB400 Bruker spectrometer and a liquid probe.*

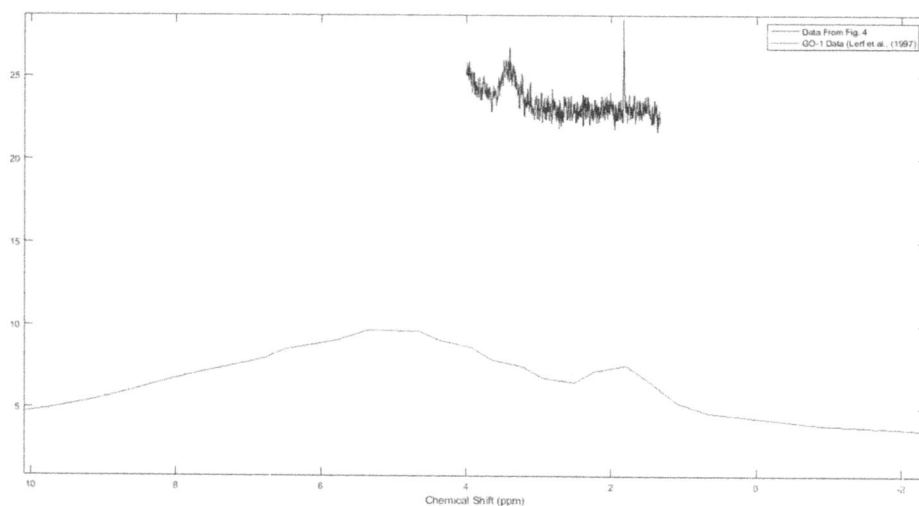

Figure 6.
*A comparison of the data in **Figure 5** with the 1H SS NMR of graphene oxide obtained by MAS as reported in [14].*

the results were compared [14]. The ^{1}H MAS data of D_2O processed graphene oxide reported in [14] exhibit peaks in almost exact locations, as seen in **Figure 5**. The data shown in **Figure 6** ensures that the suspended solids can be studied using liquid NMR spectroscopy by observing the close agreement with the MAS spectrum without any spinning.

5. NMR of adsorbed molecules using a benchtop spectrometer

The work presented so far has ensured that we can use NMR spectrometers designed for liquids to study domains of solids with sufficient mobility. Adsorbed molecules on a solid catalyst surface can have sufficient mobility to be accessible to the techniques of liquid NMR spectroscopy. An early study of ethylene adsorbed on supported Ru-Cu bimetallic catalyst revealed that part of the spectrum did not benefit from the use of solid NMR techniques. Instead, high-resolution liquid NMR techniques were needed to identify the spectral characteristics of the observed species [11]. Furthermore, selective saturation data during *in situ* ^{1}H NMR of adsorbed hydrogen over Ru/SiO_2 revealed a rapid exchange of adsorbed hydrogen with the gas phase [20]. As such, NMR under operando conditions can enable us to monitor the mobility of the molecules and the impact of this mobility on their reactivity during catalysis. Standard measurements such as single pulse spectroscopy and traditional multiple pulses for relaxation time measurements can produce a wealth of information about the adsorbate molecules and how they interact with their environment, including the support surface and at the metal-support interface.

Currently, benchtop systems provide a different configurational advantage of having a close-coupled gas manifold. A manifold that is closely coupled to the spectrometer enables low volumes of gas such that it is possible to collect volumetric information along with the NMR data and have a rapid transfer of gas to the spectrometer without too much delay, ensuring that it is possible to monitor surface processes in real time. The disadvantages of resolution are overcome by the broad lines induced by the surface roughness, and when working with metals, large shifts that are due to electron-rich or deficient surfaces.

In our lab, we employed a 1 Tesla MAGRITEK NMR spectrometer to investigate the adsorbed molecules. We chose to study supported Pd (palladium) as our initial metal, due to its very low reduction temperature. Furthermore, β-PdH is a widely studied structure through NMR spectroscopy in the literature [24, 25]. Our initial attempts were immediately successful as we were able to confirm both the adsorption isotherm and quantify the NMR spectra.

H_2 chemisorption over %1.66 wt. Pd/Ceria catalyst was monitored through ^{1}H NMR spectra. For *in situ* measurements, the NMR sample tube was connected by vacuum connectors to the high vacuum manifold, which can sustain pressures better than 10^{-5} Torr (**Figure 7**). The sample was exposed to different hydrogen pressures between 10 and 700 Torr while collecting ^{1}H-NMR spectra at each pressure. Before starting the experiment, the sample was exposed to 150 Torr H_2 pressure to reduce the sample to ambient temperature to change state from metal oxide to metal overnight. After reducing the sample, it was exposed to a vacuum to remove water from the surface. Then, the sample was exposed to different hydrogen pressures, which are 12.9, 23.0, 42.1, 53.6, 104.5, 203.2, 309.6, 411.0, 511.2, 620.5, and 711.7 Torr (valve that is placed before the NMR tube was closed). At the beginning, the experiment

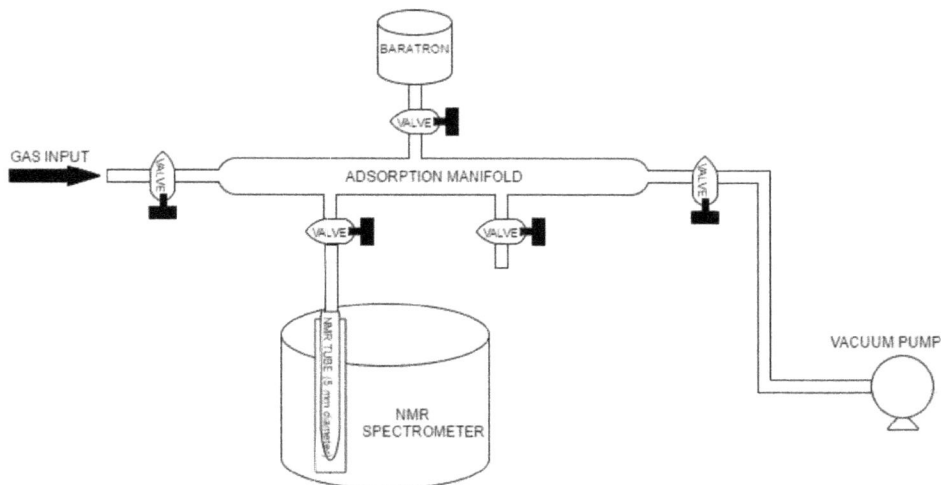

Figure 7.
Schematic representation of experimental set-up.

was carried out at 12.9 Torr H_2 pressure. To perform the experiment at 12.9 Torr, the hydrogen was sent from the gas input by opening the valve, and the hydrogen pressure in the manifold was read using the Baratron. Then, the valve which is placed above the NMR tube was opened, and the hydrogen pressure started to decrease until reaching equilibrium because of adsorption. From opening the valve that is placed above the NMR tube to reaching equilibrium pressure, the NMR spectrometer was run in hydrogen-proton + mode to record ^{1}H-NMR spectra. After the system reached equilibrium (9.9 Torr), the pressure was increased to 23.0 Torr H_2 pressure. For all pressures, ^{1}H-NMR spectra were recorded separately. This procedure was followed until reaching 711.7 Torr H_2 pressure. These were performed at room temperature because %1.66 wt. Pd/Ceria can be reduced at room temperature. Then, dead volume measurement was carried out using He to determine the volume of the system. Finally, to determine ceria's hydrogen peak frequency, ceria was examined with the NMR spectrometer overnight.

A peak detected at ~5 ppm without any hydrogen treatment was assigned to the OH groups on ceria surface. First exposure to hydrogen reduced Pd *in situ*, followed by evacuation. Subsequent doses of hydrogen revealed both an adsorption over Pd and a distinct peak for the PdH formation. After 20 Torr H_2 pressure, a second peak at ~25 ppm of the ^{1}H-NMR spectrum was observed and assigned to the β phase of palladium hydride (**Figure 8**). In order to reveal the quantitative comparisons, the adsorption isotherms from volumetric data are also reported in **Figure 9**. The integrated intensities, obtained from the areas under the deconvoluted peaks, were consistent with the quantified adsorption amounts from the volumetric chemisorption experiment. The chemical shift values were comparable with the reported data in the literature, demonstrating the potential of using a benchtop unit designed primarily for a liquid phase measurement toward measurements of adsorbed species [24, 26]. After determining areas under the peaks, the adsorbed amounts were determined by pressure measurements, and the adsorbed amount (H/Pd ratio) at the equilibrium point, which is at 184.60 Torr H_2 pressure, was found as 0.59, and the adsorbed amounts were consistent with the NMR peak areas.

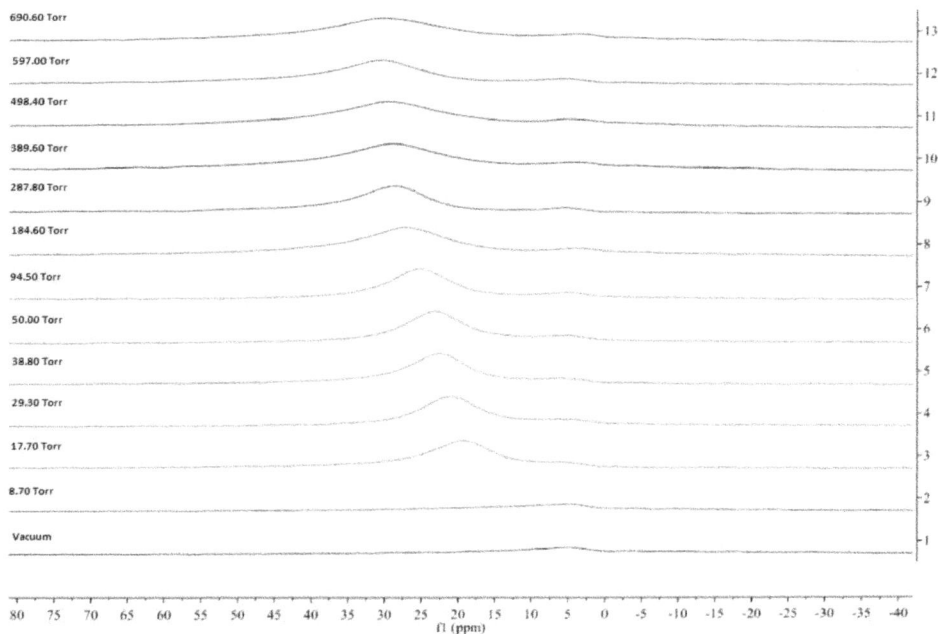

Figure 8.
Stacked NMR spectra exhibiting the pressure dependency of the signals emanating from CeO$_2$ as well as βPdH.

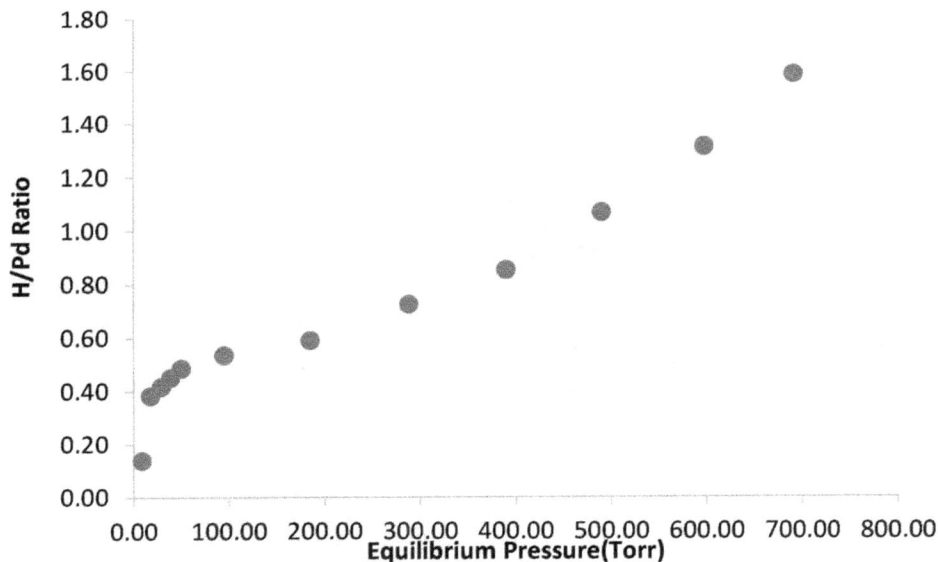

Figure 9.
H/Pd ratios of the adsorption isotherm obtained from volumetric measurements.

The comparison of the total integrated intensities with the volumetric data is given in **Figure 10**. The semi-quantitative agreement in the data shown in **Figure 10** was further clarified in later studies [27].

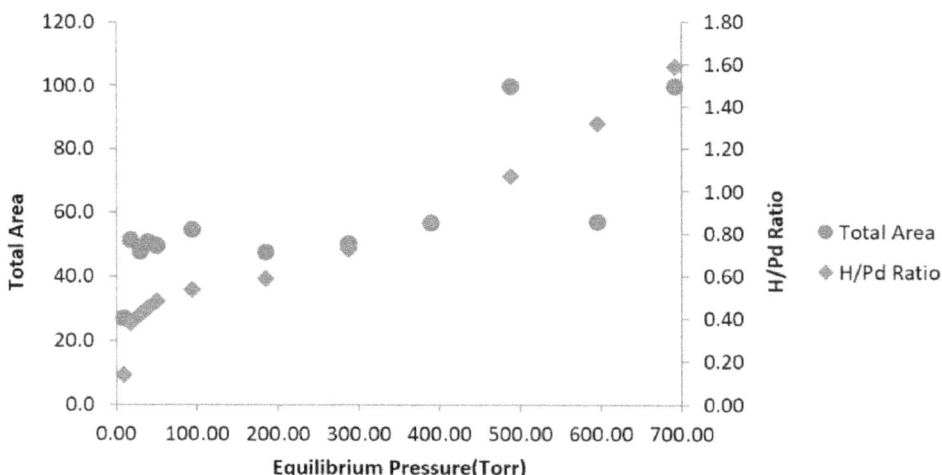

Figure 10.
The comparison between the isotherm and the total integrated intensity of the spectrum.

After collecting NMR data, which is quantitatively in agreement with the volumetric information, a number of studies were performed, which will be summarized here, and the reader is referred to their original publications. The role of Pd as a reduction promoter of TiO_2 was thoroughly investigated using temperature-programmed reduction and NMR spectroscopy [28]. The TPR data were quantitatively analyzed, and domains of partially reduced TiO_2 were identified. Furthermore, the TPR data were interpreted in terms of the amount of the βPdH available. The stoichiometry allowed us to surmise the possibility of finding 2D Pd domains on the surface, which was correlated with the Pd loading. At low Pd loadings, the βPdH stoichiometries were lower than the known equilibrium values of 0.60. In addition, the presence of 2D Pd domains was confirmed through TEM measurements. The presence of these 2D domains and the reducibility of the TiO_2 were strongly correlated and quantified.

In another study, DFT NMR studies revealed the distinct features assigned to αPdH to a specific hydrogen state that is adsorbed between a surface PdO layer and a bulk metallic Pd. The electronic states of these hydrides and their strong dependence on hydrogen pressure were also identified. The existence of a dynamic oxidation state is indicated as the hydrogen partial pressures increase, shifting the adsorbed state from αPdH to βPdH; their role in chemical reactions is under investigation in the present studies [27].

6. Conclusions

The dynamics of the adsorbed species play a significant role in the activity and the selectivity during catalytic reactions. NMR spectroscopy stands out among the operando studies in terms of the information that can be gathered about the motional characteristics, along with the structural details. This chapter targeted convincing the readers that using low-field liquid spectrometers could be used to study the adsorbed molecules. The limitation so far is the inability to perform temperature control, which is being investigated in our labs and elsewhere.

Acknowledgements

The ability to use a liquid NMR spectrometer to study adsorbed molecules was first demonstrated during a sabbatical visit to the UC Berkeley Chemical and Biomolecular Engineering Department. Prof. Jeffrey A. Reimer and Prof. Enrique Iglesia are kindly acknowledged for their generous pledges of their time and laboratories.

Measurements performed in Berkeley College of Chemistry NMR facilities (current Pines Magnetic Resonance Center) could not have been possible without the technical skills, professional support, and assistance of its director, Dr. Hasan Celik. His collegiality as well as friendship are kindly acknowledged.

DU is grateful for the grants provided by the TUBITAK National Leader Researcher Program, under grant no 120C150, for the studies performed demonstrating the use of a benchtop NMR spectrometer to study adsorbed molecules.

Conflict of interest

The authors declare no conflict of interest.

Author details

Deniz Uner* and Ecem (Volkan) Deveci
Chemical Engineering, Middle East Technical University, Ankara, Turkey

*Address all correspondence to: uner@metu.edu.tr

IntechOpen

References

[1] Ben TY, Fraissard J. Nuclear magnetic resonance in heterogeneous catalysis. In: Imelik B, Vedrine JC, editors. Catalyst Characterization: Physical Techniques for Solid Materials. New York: Plenum Press; 1994. pp. 91-129

[2] Andrew ER, Newing RA. The narrowing of nuclear magnetic resonance spectra by molecular rotation in solids. Proceedings of the Physical Society of London. 1958;**72**(468):959-972

[3] Andrew E, Bradbury A, Eades R. Nuclear magnetic resonance spectra from a crystal rotated at high speed. Nature. 1958;**182**:1659-1659

[4] Andrew ER, Farnell LF. The effect of macroscopic rotation on anistotropic bilinear spin interactions in solids. Molecular Physics. 1968;**15**(2):157-165

[5] Blasco T. Insights into reaction mechanisms in heterogeneous catalysis revealed by in situ NMR spectroscopy. Chemical Society Reviews. 2010;**39**(12):4685-4702. Available from: https://pubs.rsc.org/en/content/articlehtml/2010/cs/c0cs00033g

[6] Zhang W, Xu S, Han X, Bao X. In situ solid-state NMR for heterogeneous catalysis: A joint experimental and theoretical approach. Chemical Society Reviews. 2011;**41**(1):192-210. Available from: https://pubs.rsc.org/en/content/articlehtml/2012/cs/c1cs15009j

[7] Hunger M. In situ flow MAS NMR spectroscopy: State of the art and applications in heterogeneous catalysis. Progress in Nuclear Magnetic Resonance Spectroscopy. 2008;**53**(3):105-127

[8] Uner DO, Pruski M, King TS. The role of alkali promoters in Fischer-Tropsch synthesis on Ru/SiO$_2$ surfaces. Topics in Catalysis. 1995;**2**(1-4):59-69

[9] Uner DO, Pruski M, Gerstein BC, King TS. Hydrogen chemisorption on potassium promoted supported ruthenium catalysts. Journal of Catalysis. 1994;**146**(2):530-536. DOI: 10.1006/jcat.1994.1091

[10] Hwang SJ, Uner DO, King TS, Pruski M, Gerstein BC. Characterization of silica catalyst supports by single and multiple quantum proton NMR spectroscopy. Journal of Physical Chemistry. 1995;**99**(11):3697-3703

[11] Pruski M, Kelzenberg JC, Sprock M, Gerstein BC, King TS. 13C High resolution NMR studies of adsorption and reaction of ethylene on silica-supported Ru and RuCu catalysts. Colloids and Surfaces. 1990;**45**(C):39-56

[12] Pruski M, Kelzenberg JC, Gerstein BC, King TS. Solid-state nmr of 13c in ethylene adsorbed on silica-supported ruthenium. Journal of the American Chemical Society. 1990;**112**(11):4232-4240. Available from: https://pubs.acs.org/doi/abs/10.1021/ja00167a019

[13] Klinowski J, Solid-State NMR. Studies of molecular sieve catalysts. Chemical Reviews. 1991;**91**:1459-1479. Available from: https://pubs.acs.org/sharingguidelines

[14] Lerf A, He H, Riedl T, Forster M, Klinowski J. 13C and 1H MAS NMR studies of graphite oxide and its chemically modified derivatives. Solid State Ionics. 1997;**101-103**(PART 2):857-862

[15] Hunger M. In situ NMR spectroscopy in heterogeneous catalysis. Catalysis Today. 2004;**97**(1):3-12

[16] Engelke F, Vincent R, King TS, Pruski M. Adsorption, desorption, and interparticle motion of hydrogen on silica-supported ruthenium: A study by in situ nuclear magnetic resonance. The Journal of Chemical Physics. 1994;**101**(9):7262-7272. Available from: http://aip.scitation.org/doi/10.1063/1.468497

[17] Fukushima E, Roeder SBW. Experimental Pulse NMR. Boca Raton, FL: CRC Press; 2018

[18] Ernst RR, Bodenhausen G, Wokaun A. Principles of Nuclear Magnetic Resonance in One and Two Dimensions. Oxford: Oxford University Press; 1990

[19] Sachleben JR, Wooten EW, Emsley L, Pines A, Colvin VL, Alivisatos AP. NMR studies of the surface structure and dynamics of semiconductor nanocrystals. Chemical Physics Letters. 1992;**198**(5):431-436

[20] Engelke F, Bhatia S, King TS, Pruski M. Dynamics of hydrogen at the surface of supported ruthenium. Physical Review B. 1994;**49**(4):2730-2738

[21] Uner DO, Savargoankar N, Pruski M, King TS. The effects of alkali promoters on the dynamics of hydrogen chemisorption and syngas reaction kinetics on Ru/SiO$_2$ surfaces. In: Froment GF, Waugh KC, editors. Dynamics of Surfaces and Reaction Kinetics in Heterogeneous Catalysis Proceedings of the International Symposium, Studies in Surface Science and Catalysis. Vol. 109. Elsevier; 1997. pp. 315-324. Available from: http://www.sciencedirect.com/science/article/pii/S0167299197804181

[22] NIST. 2025. Available from: https://www.nist.gov/publications/composition-determination-low-pressure-gas-phase-mixtures-1h-nmr-spectroscopy

[23] Bowers GM, Schaef HT, Miller QRS, Walter ED, Burton SD, Hoyt DW, et al. 13C nuclear magnetic resonance spectroscopy of methane and carbon dioxide in a natural shale. ACS Earth and Space Chemistry. 2019;**3**(3):324-328. Available from: https://pubs.acs.org/doi/full/10.1021/acsearthspacechem.8b00214

[24] Barabino DJ, Dybowski C. Nuclear magnetic resonance of hydrogen sorbed by powdered palladium metal and alumina-supported palladium. Solid State Nuclear Magnetic Resonance. 1992;**1**:5-12

[25] Hanneken JW, Baker DB, Conradi MS, Eastman JA. NMR study of the nanocrystalline palladium–hydrogen system. Journal of Alloys and Compounds. 2002;**330-332**:714-717

[26] Frieske H, Wicke E. Magnetic susceptibility and equilibrium diagram of PdH. Berichte der Bunsengesellschaft für Physikalische Chemie. 1973;**77**(1):48-52

[27] Mete E, Yilmaz B, Uner D. PdH α-phase is associated with residual oxygen as revealed by in situ 1H NMR measurements and DFT-NMR estimations. Applied Surface Science. 2023;**641**:158421

[28] Yarar M, Bouziani A, Uner D. Pd as a reduction promoter for TiO$_2$: Oxygen and hydrogen transport at 2D and 3D Pd interfaces with TiO$_2$ monitored by TPR, operando 1H NMR and CO oxidation studies. Catalysis Communications. 2023;**174**:106580

Chapter 4

Chemical Exchange Dynamics and Variable Frequency CPMG Echo Train

Aleš Mohorič

Abstract

Transversal relaxation in Nuclear Magnetic Resonance (NMR) spectroscopy is a fundamental process that influences the decay of magnetization in the plane perpendicular to the static magnetic field. This chapter aims to provide an analysis of transversal relaxation, encompassing its theoretical background, practical applications, and recent advancements, specifically focusing on its utility in probing the frequency-dependent spectral characteristics of chemical exchange processes. The study of NMR relaxation offers a look at chemical exchange dynamics and provides unique insights into molecular kinetics and conformational changes in solution-phase systems. Carr-Purcell-Meiboom-Gill (CPMG) methodologies have implications for broader chemical and biological studies, along with potential technological advancements to enhance spectral analysis. The integration of variable frequency CPMG techniques into the study of chemical exchange kinetics represents a critical advancement in NMR spectroscopy. This approach not only extends the range of measurable exchange rates but also reveals intricate details about intermediate states in chemical reactions.

Keywords: exchange rate, CPMG sequence, relaxation rate, NMR, exchange rate spectrum

1. Introduction

The study of chemical exchange dynamics through Nuclear Magnetic Resonance (NMR) provides unique insights into molecular kinetics and conformational changes in solution-phase systems. This chapter outlines the application and advancements of the Carr-Purcell-Meiboom-Gill (CPMG) NMR method, specifically focusing on its utility in probing the frequency-dependent spectral characteristics of chemical exchange processes. By leveraging the findings from two recent studies, this chapter will elucidate the underlying principles of the variable frequency CPMG echo train and its role in capturing the complex interplay of molecular motion and exchange kinetics.

NMR spectroscopy has been pivotal in elucidating the structure, dynamics, and interactions of molecules across various scientific disciplines. Transversal relaxation is central to understanding NMR signal decay and spectral line broadening, which directly impact the resolution and sensitivity of NMR experiments. Bloch's pioneering work [1] introduced the concept of nuclear induction and formulated the basic principles governing the behavior of nuclear spins in a magnetic field. Bloembergen, Purcell, and Pound [2] extended Bloch's work by exploring relaxation phenomena in NMR and described how they depend on molecular motion and interactions, including chemical exchange. Carr and Purcell introduced the Carr-Purcell sequence [3], which is essential for mitigating dephasing due to magnetic field inhomogeneities. Their work showed that applying a train of π pulses could refocus dephasing spins, enabling accurate measurements of T_2. By minimizing artifacts caused by diffusion and field inhomogeneities, their technique allows to better isolate dynamic exchange phenomena in complex systems. Meiboom and Gill [4] modified the original Carr-Purcell sequence to create a more robust method—now known as the CPMG sequence—for measuring T_2 relaxation times. By introducing phase cycling to correct imperfections in the π pulses, we can significantly improve the reliability of spin echo measurements. Loria and co-workers developed a modified CPMG sequence desig-ned specifically to compensate for relaxation effects during the measurement of chemical exchange [5]. By minimizing artifacts due to relaxation, their method enabled a clearer separation of exchange contributions from intrinsic T_2 relaxation. Redfield [6] developed a comprehensive theory of relaxation based on quantum mechanics. McConnell intro-duced the idea that chemical exchange between different molecular environments can be measured by NMR [7]. He derived equations that relate exchange rates to changes in NMR line shapes. His work demonstrated that the exchange process alters the effective relaxation rates observed in an NMR experiment. Carver and Richards [8] extended earlier work by providing a quantitative description of how chemical exchange affects NMR relaxation. Luz and Meiboom investigated the effects of proton chemical exchange on NMR line shapes and relaxation rates [9]. Their work provided one of the first detailed accounts of how rapid exchange processes influence observed NMR signals, particularly in systems where exchange rates are comparable to the frequency differences between sites. Forsén and Hoffman [10] contributed to the early understanding of spin relaxation by exploring the mechanisms of cross-relaxation in simple liquids. Their experiments elucidate the role of dipole-dipole interactions in driving relaxation processes and provide insights into how molecular motion influences NMR line shapes. Kay and colleagues [11] demonstrated how 15 N NMR can be used to probe the backbone dynamics of proteins. Their experiments revealed how chemical exchange processes on the microsecond-to-millisecond timescale affect relaxation rates, providing important insights into protein conformational dynamics. The methodology they introduced has been widely adopted for studying biomolecular dynamics, particularly using CPMG relaxation dispersion tech-niques. Tollinger and colleagues applied CPMG relaxation dispersion techniques to study slow dynamic processes in proteins [12]. Their work provided detailed insights into the conformational exchange between folded and unfolded states, linking dynamic behavior to function. Clore and Gronenborn provided one of the early comprehensive reviews on using NMR to study protein dynamics, including chemical exchange effects [13]. Palmer reviews methods for investigating slow dynamics in biological macromolecules using NMR in [14, 15]. He discusses the theoretical basis for relaxation dispersion experiments and the impact of chemical exchange on NMR line shapes.

2. Theory

2.1 Bloch equations

NMR is a spectroscopic technique that leverages the magnetic properties of atomic nuclei to elucidate the structure, dynamics, and chemical environment of molecules.

Certain nuclei possess an intrinsic quantum mechanical property known as spin. The spin quantum number, I, can be an integer or half-integer. Spinning nuclei generate a magnetic moment (μ) given by $\mu = \gamma \hbar I$, where γ is the nucleus-specific gyromagnetic ratio. When placed in a strong, uniform magnetic field B_0, the nuclear spins align parallel or antiparallel to the field direction. The energy difference between these states is $\Delta E = \gamma \hbar B_0$. This energy difference corresponds to the frequency at which the nucleus can absorb or emit electromagnetic radiation, known as the Larmor frequency:

$$\omega = \gamma B_0. \tag{1}$$

The nuclear magnetic moment does not merely align with B_0 but precesses around it like a spinning top. This precession is essential for NMR signal detection. The Bloch equations describe the time evolution of the macroscopic magnetization in a spin system under the influence of an external magnetic field and relaxation processes. If we consider a system whose state is represented by a density matrix ρ in Hilbert space, the time evolution of ρ is governed by the Liouville-von Neumann equation,

$$\frac{d\rho}{dt} = -\frac{i}{\hbar}[\mathcal{H}, \rho], \tag{2}$$

where $\mathcal{H}$ is the Hamiltonian operator of the system. For a spin in an external magnetic field $\mathbf{B}$, the Hamiltonian is given by

$$\mathcal{H} = -\gamma \hbar \hat{\mathbf{I}} \cdot \mathbf{B}, \tag{3}$$

with $\hat{\mathbf{I}} = (\hat{I}_x, \hat{I}_y, \hat{I}_z)$ the spin operator vector. The observable magnetization $\mathbf{M}(t)$ is defined as the expectation value of the spin operator:

$$\mathbf{M}(t) = \mathrm{Tr}(\rho \hat{\mathbf{I}}). \tag{4}$$

To derive the equation of motion for the magnetization, we substitute the Liouville-von Neumann equation in $\dot{\mathbf{M}} = \mathrm{Tr}(\dot{\rho}\hat{\mathbf{I}})$ to get $\frac{d\mathbf{M}}{dt} = -\frac{i}{\hbar}\mathrm{Tr}([\mathcal{H}, \rho]\hat{\mathbf{I}})$. Exploiting the cyclic invariance of the trace $\mathrm{Tr}([\mathcal{H}, \rho]\hat{\mathbf{I}}) = \mathrm{Tr}(\rho[\hat{\mathbf{I}}, \mathcal{H}])$, and given the Hamiltonian Eq. 16, the commutator $[\hat{\mathbf{I}}, \mathcal{H}]$ can be calculated by first noting that $[\hat{\mathbf{I}}, \hat{\mathbf{I}} \cdot \mathbf{B}] = \sum_i B_i [\hat{\mathbf{I}}, \hat{I}_i]$. Recalling the fundamental commutation relations for angular momentum, $[\hat{I}_i, \hat{I}_j] = i\hbar \, \varepsilon_{ijk}\hat{I}_k$, we can show that $[\hat{\mathbf{I}}, \mathcal{H}] = \gamma \hbar^2 \hat{\mathbf{I}} \times \mathbf{B}$. Thus, the time evolution of the magnetization becomes

$$\frac{d\mathbf{M}}{dt} = \gamma \mathrm{Tr}(\rho \hat{\mathbf{I}} \times \mathbf{B}) = \gamma \mathbf{M} \times \mathbf{B}. \tag{5}$$

2.2 Relaxation

This last equation represents the coherent, unitary evolution of the spin under the Zeeman interaction: the magnetization vector precesses about the magnetic field at an angular frequency γB. While the above derivation captures the reversible, coherent dynamics, real spin systems are open and interact with their environments. Such interactions cause dissipation and decoherence, which are phenomenologically described by relaxation times. Longitudinal relaxation T_1 characterizes the return of the magnetization component along the field to its thermal equilibrium value M_0. Transverse relaxation T_2 describes the decay of magnetization components perpendicular to the field due to dephasing. This spin behavior is described by modified Bloch equations:

$$\frac{dM_x}{dt} = \gamma(M_y B_z - M_z B_y) - \frac{M_x}{T_2},$$
$$\frac{dM_y}{dt} = \gamma(M_z B_x - M_x B_z) - \frac{M_y}{T_2}, \tag{6}$$
$$\frac{dM_z}{dt} = \gamma(M_x B_y - M_y B_x) - \frac{M_z - M_0}{T_1}.$$

Transversal relaxation refers to the process by which the coherent magnetization in the transverse plane decays due to dephasing among individual spins. Following an RF excitation, spins precess at the Larmor frequency about the static magnetic field. In an idealized, perfectly homogeneous field, these spins would precess in unison, maintaining phase coherence and a robust transverse signal. In practice, however, local magnetic field fluctuations arise from intrinsic interactions—such as dipole-dipole couplings—and extrinsic factors like field inhomogeneities and sample imperfections. These variations cause spins to precess at slightly different frequencies, leading to a loss of phase coherence. Transversal relaxation is linked to the decay of off-diagonal elements of the density matrix because of interactions such as spin-spin couplings and local field fluctuations. A microscopic treatment relates T_2 relaxation to the spectral density of these fluctuations evaluated at the Larmor frequency. In this picture, the relaxation process is inherently probabilistic and results from the ensemble averaging over many small, random interactions that shift the individual spin precession frequencies.

In many complex systems, the decay of the transverse magnetization is not strictly a single exponential. Anisotropic interactions, including chemical shift anisotropy and quadrupolar couplings, can lead to non-exponential or multi-exponential decay behavior. Furthermore, chemical exchange processes, in which nuclei rapidly interchange between distinct environments, add another layer of complexity by introducing additional dephasing channels.

The free induction decay (FID) signal is the observable consequence of the decay of transverse magnetization following an RF excitation pulse. Typically, a spin system interacts weakly with its environment—also referred to as the bath—and the full Hamiltonian is given by

$$\mathcal{H} = \mathcal{H}_0 + \mathcal{H}_{\text{int}} + \mathcal{H}_{\text{bath}}. \tag{7}$$

Here, $\mathcal{H}_{\text{int}}$ represents the interaction between the spins and the bath degrees of freedom. Within the weak-coupling (Born) and Markov approximations, the evolution of the reduced density matrix $\rho(t)$ (tracing out the bath variables) can be described by a quantum master equation of the form

$$\frac{d\rho}{dt} = -\frac{i}{\hbar}[\mathcal{H}_0, \rho] + \mathcal{R}[\rho], \tag{8}$$

where $\mathcal{R}[\rho]$ is the Redfield relaxation superoperator that accounts for dissipative effects [16]. The Redfield theory provides a microscopic derivation of relaxation rates by relating them to the spectral densities of the bath fluctuations. In the interaction picture, one finds that the coherences evolve as

$$\rho_{ij}(t) \propto \rho_{ij}(0) e^{-i\omega_{ij}t} e^{-R_{ij}t}, \tag{9}$$

where ω_{ij} is the energy difference between states i and j, and R_{ij} is the relaxation rate associated with dephasing. In many cases, one can identify $2R_{ij} \approx 1/T_2$. The decay constant T_2 is determined by the Fourier transform of the autocorrelation function of the fluctuating fields:

$$\frac{1}{T_2} \sim J(\omega_0), \tag{10}$$

where $J(\omega)$ denotes the spectral density at the Larmor frequency ω_0. Thus, T_2 relaxation is a direct manifestation of the environmental fluctuations that randomly shift the precession frequencies of the spins.

The observable FID signal is proportional to the macroscopic transverse magnetization $M_\perp(t)$, which, in the quantum picture, is given by

$$M_\perp(t) = \mathrm{Tr}\left[\rho(t)\left(\hat{I}_x + i\hat{I}_y\right)\right]. \tag{11}$$

After an RF pulse, the system is brought into a state with significant transverse coherence. As time progresses, the decay of the off-diagonal elements leads to a signal of the form

$$M_\perp(t) = M_\perp(0) e^{-t/T_2} e^{-i\omega_0 t}. \tag{12}$$

Here, the exponential e^{-t/T_2} arises directly from the decay of the coherences due to the bath-induced fluctuations. The phase factor $e^{-i\omega_0 t}$ reflects the Larmor precession of the spins. The precise form of the FID signal can be derived by solving the master equation for $\rho(t)$ under the appropriate initial conditions and then computing the trace with the transverse spin operator. A key quantity in the quantum description of T_2 relaxation is the spectral density $J(\omega)$ of the fluctuating magnetic fields. This function is defined as the Fourier transform of the time-correlation function of the environmental perturbations:

$$J(\omega) = \int_{-\infty}^{+\infty} \langle B_{\mathrm{fluct}}(t) B_{\mathrm{fluct}}(0) \rangle e^{-i\omega t} dt. \tag{13}$$

For a Markovian environment, the spectral density is often approximated as a Lorentzian function, and its value at ω_0 determines the rate at which phase coherence is lost. By measuring the FID decay, one can extract T_2 and, in turn, gain insights into the dynamics of the local environment influencing the spins.

While the intrinsic quantum mechanical T_2 relaxation arises from spin-spin interactions and environmental fluctuations, practical NMR experiments often encounter additional dephasing due to static magnetic field inhomogeneities. These extrinsic effects lead to an effective relaxation time T_2^* that is shorter than the intrinsic T_2. Spin echoes, are used to refocus these inhomogeneities and reveal the true T_2 relaxation.

Moreover, in complex systems where spins undergo chemical exchange between different environments, the dephasing mechanism can become non-exponential.

Quantum mechanical models that incorporate exchange processes lead to modified master equations with additional terms that couple different spin environments. These models provide a deeper understanding of the microscopic mechanisms that influence the decay of the FID signal.

2.3 Chemical shift

While the Bloch equations and transversal relaxation provide insights into the overall behavior of nuclear spins in a magnetic field, two critical interactions—chemical shift and J-coupling—offer even deeper structural information. Chemical shift arises from the local electronic environment that shields the nuclei from the external magnetic field, whereas J-coupling (or scalar coupling) originates from indirect interactions between nuclear spins mediated by bonding electrons.

In an external magnetic field $\mathbf{B}_0$, a bare nucleus with a magnetic moment μ would experience a Zeeman interaction described by the Hamiltonian $\mathcal{H} = -\mu \cdot \mathbf{B}_0$. However, in a molecule the nucleus is surrounded by electrons that generate local magnetic fields and induce a screening effect that partially opposes $\mathbf{B}_0$, modifying the effective field experienced by the nucleus. The net field is given by $\mathbf{B}_{\text{eff}} = \mathbf{B}_0(1 - \sigma)$, where σ is the shielding constant. The Hamiltonian for the nuclear Zeeman interaction then becomes

$$\mathcal{H}_{cs} = -\gamma \hbar \hat{I}_z B_0 (1 - \sigma), \tag{14}$$

with $\hat{I}_z$ the z-component of the nuclear spin operator.

While chemical shift modifies the resonance frequency of a nucleus, J-coupling arises from the indirect interaction between nuclear spins mediated by bonding electrons. This interaction is isotropic (in liquids, where rapid molecular tumbling averages out anisotropic effects) and is typically described by the Hamiltonian

$$\mathcal{H}_J = 2\pi J_{ij} \hat{\mathbf{I}}_i \cdot \hat{\mathbf{I}}_j, \tag{15}$$

where J_{ij} is the scalar coupling constant between spins i and j, and $\hat{\mathbf{I}}_i$ and $\hat{\mathbf{I}}_j$ are the corresponding nuclear spin operators. The coupling constant J_{ij} is determined by the overlap of electronic wavefunctions that mediate the interaction.

In a realistic NMR experiment, both chemical shift and J-coupling affect the nuclear spin system. The combined spin Hamiltonian for a two-spin system can be written as

$$\mathcal{H} = -\gamma_i \hbar (1 - \sigma_i) \hat{I}_{i,z} B_0 - \gamma_j \hbar (1 - \sigma_j) \hat{I}_{j,z} B_0 + 2\pi J_{ij} \hat{\mathbf{I}}_i \cdot \hat{\mathbf{I}}_j. \tag{16}$$

Here, the first two terms represent the modified Zeeman interactions (chemical shifts) for nuclei i and j, and the third term represents the scalar coupling between them. In an NMR experiment, the relative magnitudes of these terms determine the spectral features. When the chemical shift difference is large compared to J_{ij}, the multiplet splitting is straightforward. However, in cases of nearly equivalent nuclei (small chemical shift difference), complex coupling patterns may emerge due to level mixing and second-order effects.

In the density matrix formalism, the initial density matrix after the RF pulse contains both populations and coherences. The evolution of the system is governed by the master equation (Eq. 8), where $\mathcal{H}_0$ is replaced by $\mathcal{H}$ (Eq. 16) that includes both the chemical shift and J-coupling terms. The off-diagonal elements of ρ (coherences) evolve under the combined influence of the different precession frequencies (due to chemical shift) and the coupling-induced splittings. In many systems, molecular motions modulate both the chemical shift and the J-coupling. Rapid exchange or conformational fluctuations can average anisotropic interactions, leading to effective isotropic couplings in solution.

2.4 CPMG

The CPMG sequence is a cornerstone technique for measuring T_2 relaxation times and for refocusing dephasing effects due to static magnetic field inhomogeneities. The spin echo sequence, introduced by Hahn, uses a single π pulse applied at time τ after an initial excitation pulse to refocus dephasing caused by static inhomogeneities. The density matrix $\rho(t)$ evolves under the Hamiltonian $\mathcal{H}_0$ so that off-diagonal coherences decay due to frequency dispersion. A single π pulse ideally reverses the phase evolution, leading to an echo at time 2τ. However, in many practical scenarios—including systems with pulse imperfections or evolving spin–spin interactions—a single echo may not suffice. The CPMG sequence improves upon this by applying a series of π pulses to generate multiple echoes, thereby averaging out both static and slowly varying inhomogeneities. After the initial $\pi/2$ pulse, the system is prepared in a coherent superposition of spin states. Under free evolution, the density operator in the laboratory frame evolves as

$$\rho(t) = U(t)\rho(0)U^\dagger(t), \tag{17}$$

with the propagator $U(t) = \exp(-\frac{i}{\hbar}\mathcal{H}_0 t)$. In the presence of an inhomogeneous field, different spins precess at slightly different frequencies, leading to the dephasing of the off-diagonal elements in $\rho(t)$. When a π pulse is applied, its effect is described by a unitary operator U_π. For an ideal π rotation about the x-axis, this operator is given by $U_\pi = \exp(-i\pi \hat{I}_x)$.

After the π pulse, the evolution operator becomes

$$U_{\text{echo}} = U(\tau)U_\pi U(\tau). \tag{18}$$

In an ideal case, U_π effectively reverses the phase accrued during the first τ period, leading to rephasing at time 2τ.

To analyze the effect of a train of π pulses in the CPMG sequence, it is convenient to work in the toggling frame. In this rotating reference frame, the effect of each π pulse is to flip the sign of the Hamiltonian governing the free evolution. Let $f(t)$ be a modulation function that alternates between $+1$ and -1 depending on whether the system is in the "normal" or "inverted" phase. The effective Hamiltonian in the toggling frame is then given by $\mathcal{H}_{\text{eff}}(t) = f(t)\mathcal{H}_0$. If $f(t)$ is an even function with respect to the center of each cycle, the effects of static dephasing are canceled, leading to a reduced effective dephasing rate and a more accurate measurement of the intrinsic T_2 relaxation. Average Hamiltonian theory, which relies on a Magnus expansion, allows one to account for higher-order corrections arising from finite pulse durations and pulse imperfections. For example, the first-order correction cancels for symmetric pulse sequences, while higher-order terms introduce systematic errors that can be minimized through proper pulse design. In the original Carr-Purcell sequence, imperfections in the π pulses (e.g., slight over- or under-rotations or off-resonance effects) can accumulate and lead to an effective decay of the echo amplitudes. The Meiboom-Gill modification introduces a phase shift in the π pulses relative to the initial $\pi/2$ pulse, typically by applying the π pulses along the y-axis instead of the x-axis. This adjustment means that the unitary operator for the refocusing pulse becomes

$$U_\pi^{(y)} = \exp\left(-i\pi\hat{I}_y\right), \tag{19}$$

which helps to preserve the transverse magnetization components even in the presence of systematic errors. The full evolution under the CPMG sequence involves alternating periods of free evolution and instantaneous pulses. The overall propagator after N cycles can be written as

$$U_{\text{CPMG}}(T) = \prod_{k=1}^{N}[U(\tau)U_\pi U(\tau)], \tag{20}$$

with $T = 2N\tau$. The density matrix after the entire sequence is

$$\rho(T) = U_{\text{CPMG}}(T)\rho(0)U_{\text{CPMG}}^\dagger(T). \tag{21}$$

Because each π pulse reverses the phase evolution from static inhomogeneities, the accumulated dephasing due to these effects is largely canceled. The remaining decay in the echo amplitudes is then dominated by intrinsic relaxation processes and residual pulse imperfections.

3. Chemical exchange and relaxation

In many NMR experiments, the measured transverse relaxation rate $R_2 = 1/T_2$ is not solely determined by intrinsic spin–spin interactions and magnetic field inhomogeneities. Instead, chemical exchange processes—in which nuclei jump between different chemical environments—can contribute significantly to the observed dephasing. When these exchange processes occur on timescales comparable to the inverse of the chemical shift difference, they modulate the effective relaxation rates. The CPMG sequence is a powerful tool to probe such dynamics. To understand the effect, we incorporate exchange terms into the evolution of the density matrix, leading to a modified set of Bloch equations known as the McConnell-Bloch equations [1, 7]. Furthermore, the Carver-Richards equation [8] provides an expression for the effective relaxation rate observed in CPMG experiments as a function of the pulse repetition rate.

Consider a two-site exchange process in which a nucleus exists in two distinct environments, A and B, with populations p_A and p_B. The local environments have different resonance frequencies, ω_A and ω_B, giving rise to a chemical shift difference $\Delta\omega = \omega_B - \omega_A$. In the absence of exchange, each site would evolve independently according to the Zeeman Hamiltonian modified by local relaxation. However, chemical exchange couples the states, causing the off-diagonal elements of the density matrix to evolve under additional exchange terms. If M_A and M_B denote the transverse magnetization components, their evolution in the rotating frame is given by the McConnell-Bloch equations:

$$\frac{dM_A}{dt} = -(R_{2,A} + i\omega_A + k_{AB})M_A + k_{BA}M_B, \tag{22}$$

$$\frac{dM_B}{dt} = -(R_{2,B} + i\omega_B + k_{BA})M_B + k_{AB}M_A, \tag{23}$$

where $R_{2,A}$ and $R_{2,B}$ are the intrinsic transverse relaxation rates for sites A and B, and k_{AB} and k_{BA} are the rate constants for the exchange from A to B and vice versa. Solving these coupled equations yields eigenvalues that describe the observed decay rates. In many cases, the observed effective transverse relaxation rate, $R_{2,\text{eff}}$, is a weighted average that includes contributions from both intrinsic relaxation and exchange.

In a CPMG experiment in the presence of chemical exchange, the refocusing is incomplete because the exchange process itself modulates the phase evolution of the spins. The periodic refocusing modifies the effective evolution operator of the system. In the toggling frame, the free evolution intervals are interleaved with instantaneous exchange events. The observed echo amplitude at time T is determined by both intrinsic relaxation and the chemical exchange process. In the absence of exchange, the echo amplitude decays with the intrinsic rate $R_{2,0}$, whereas, in the presence of exchange, the decay $R_{2,\text{eff}}$ is faster. Its dependence on the CPMG pulse repetition rate $\omega_m = \pi/\tau$ provides insight into the exchange kinetics.

Carver and Richards [8] provided a theoretical treatment of chemical exchange effects in spin echo experiments, deriving an expression for $R_{2,\text{eff}}$ for a spin-1/2 nucleus undergoing two-site chemical exchange that accounts for the modulation of exchange contributions by the CPMG sequence:

$$R_{2,\text{eff}} = R_{2,0} + \frac{p_A p_B \Delta\omega^2 k_{\text{ex}}}{k_{\text{ex}}^2 + \omega_m^2/\pi^2}\left[1 - \frac{2\omega_m}{\pi k_{\text{ex}}}\tanh\left(\frac{\pi k_{\text{ex}}}{2\omega_m}\right)\right], \tag{24}$$

where $k_{\text{ex}} = k_{AB} + k_{BA}$ is the overall exchange rate. Equation (24) shows that the exchange contribution to $R_{2,\text{eff}}$ is dispersive with respect to the pulse repetition rate ω_m. This expression, valid in fast-to-intermediate exchange regimes ($k_{\text{ex}} \gtrsim \Delta\omega$), enables extraction of k_{ex}, $\Delta\omega$, and $p_A p_B$ from relaxation dispersion curves [17]. At low pulse frequencies (i.e., long τ), exchange effects are prominent and $R_{2,\text{eff}}$ is high. As ω_m increases, the contribution of exchange is effectively averaged out, and $R_{2,\text{eff}}$ approaches the intrinsic relaxation rate $R_{2,0}$.

In practice, the accuracy of the Carver-Richards equation depends on several factors. The equation is most accurate in the intermediate-to-fast exchange regime. In the slow exchange limit, separate resonances are observed, and different models must be applied. Imperfections in the π pulses (e.g., due to off-resonance effects or finite pulse durations) can introduce additional relaxation pathways that may need correction. In complex systems, higher-order contributions to the average Hamiltonian can affect the dispersion profile. These approaches have been widely applied in the study of chemical exchange processes in proteins, nucleic acids, and other complex systems [17, 18].

4. Chemical exchange spectrum

In previous sections, we developed a framework for understanding CPMG relaxation measurements and the role of chemical exchange in modulating transverse relaxation. In this continuation, we rephrase and extend the analysis presented in Ref. [19], with the investigation of the chemical exchange rate by directly sampling the spectral density function of exchange processes. The approach provides insights into molecular conformational dynamics beyond a single exchange rate parameter, especially in cases where multiple exchange processes contribute to the overall dynamics of a spin system. When spins experience fluctuating Larmor frequencies, due either to molecular translational motion in a magnetic field gradient or to conformational changes (chemical exchange), the instantaneous frequency shift can be written as $\Delta\omega(t) = b(t)k(t)$, where $k(t)$ is a stochastic step function representing random transitions between two (or more) chemical states, and $b(t)$ is the modulation function introduced by the repeated application of π pulses in the CPMG sequence.

The accumulated phase of a spin in a rotating frame (the toggling is incorporated in the $b(t)$ part) is given by $\varphi(t) = \int_0^t \Delta\omega(t')dt'$, and at the echo time (when the net contribution of the modulation function cancels), the ensemble average of the phase factor $\langle e^{i\varphi(t)} \rangle$ leads to the signal attenuation. By applying a cumulant expansion truncated at second order (the Gaussian approximation), one obtains an expression linking the echo attenuation to the spectral density function $S(\omega)$ of the exchange process. In effect, the CPMG sequence acts as a spectral sampling function with a characteristic modulation frequency. Thus, the N-th echo attenuation can be expressed as [19].

$$E(T) = E_0 \exp\left[-\frac{8}{\pi^2}\omega_m^2 S(\omega_m)N\tau\right], \qquad (25)$$

which directly links the measured relaxation to the underlying chemical exchange spectrum.

The approach has been demonstrated in a sucrose solution in water. The experimental design was such that only the slowly decaying water proton signal was analyzed, ensuring that the dynamics measured were dominated by chemical exchange rather than other rapid processes. The experiments revealed that as the CPMG modulation frequency increased, the observed effective relaxation rate decreased. This trend indicates that higher modulation frequencies more effectively average out the fluctuations due to chemical exchange, causing $R_{2,\text{eff}}$ to approach the intrinsic relaxation rate $R_{2,0}$.

In the echo train, two components could be distinguished: a fast-decaying component associated with protons in sucrose (or water bound to sucrose) and a slowly decaying component corresponding to free water protons. The analysis focused on the water signal, from which the chemical exchange spectrum $S(\omega)$ was extracted. Notably, the extracted spectrum did not follow a simple Lorentzian profile expected for a single exchange process; instead, it indicated the presence of multiple exchange processes. This complexity implies that the dynamics of molecular conformations, and thus chemical exchange, are more intricate than a single parameter (exchange rate) can capture.

5. Conclusions

This chapter examines the Carr-Purcell-Meiboom-Gill (CPMG) method's role in studying chemical exchange dynamics, focusing on its variable frequency echo train. CPMG's strength lies in isolating exchange effects, supported by frameworks like the

Carver-Richards equation, though multi-state systems and pulse imperfections pose challenges. Future integration with machine learning could further enhance its resolution and scope in NMR spectroscopy. We will end the chapter by mentioning some of the recent results of the research.

NMR relaxation dispersion and CPMG-based experiments proved essential for probing chemical exchange and conformational dynamics in proteins, enabling precise measurement of exchange rates and populations on microsecond-to-millisecond scales [20]. Optimized pulse sequences and robust data analysis minimize artifacts, distinguishing exchange from intrinsic relaxation and revealing insights into protein function, folding, and binding [21–30]. Advanced frequency-selective techniques enhance resolution and accuracy, shaping modern NMR studies of biomolecular dynamics [26, 31, 32]. In Ref. [19], the CPMG sequence's utility in sampling the spectral density of exchange processes was demonstrated. Varying inter-pulse delays samples the frequency domain, enabling detailed characterization of molecular conformational fluctuations and complex multi-state exchange in liquids. Recent innovations in pulse sequences, data analysis, and modeling have deepened insights into complex systems. Korzhnev et al. [33, 34] pioneered the detection of low-populated protein states, mapping energy landscapes with NMR and computational integration. Vallurupalli et al. [35, 36] linked relaxation rates to exchange parameters, aiding enzyme and allostery research. Farrow et al. [32, 37] resolved folding dynamics, bridging experiment and theory. Koss et al. [38, 39] refined CPMG for flexibility and subtle exchanges, while Massi et al. [40] tackled overlapping resonances. Clore et al. [41] and Kay et al. [42] enhanced transient dynamics detection and kinetic accuracy.

Parallel methodological advances include Hansen et al.'s [43, 44] frameworks for exchange rate precision, Loria et al.'s [45, 46] optimization guidelines, and Mulder et al.'s [47] signal overlap solutions. Palmer [48] reviewed protocols for kinetic modeling, Griesinger et al. [49] minimized artifacts, and Nietlispach [50] offered practical NMR dynamics guidance. These efforts have elevated NMR-based biomolecular research. In non-protein systems, CPMG excels in characterizing exchange. Smith et al. [51] quantified small molecule dynamics, Alaei et al. [52] probed solvent motions, and Zhang et al. [53] studied supramolecular complexes. Banyai [54], Deev et al. [55], and Wang et al. [56] analyzed coordination complexes, heterocycles, and catalysts, respectively. Anderson et al. [57], Li et al. [58], Agarwal et al. [59], and Holzgrabe et al. [60] extended CPMG to ionic liquids, polymers, nanomaterials, and pharmaceuticals, showcasing its broad kinetic utility. Recent studies also applied CPMG to multiphase flow in porous media [61], metabolomics in muscle disorders [62], and low-power sequences in liquids [63]. Variable frequency CPMG advanced diffusion studies [64, 65], T_2 measurements in serum [66], and dye detection in food [67]. Applications span wood extractives [68], casein gels [69], pore sizing [70], and sandstone wettability [71], with optimized acquisition for glasses [72] and phase cycling for porous media [73]. Further uses include 2D relaxation spectroscopy [74], milk fat analysis [75], spectral editing [76], soil humin [77], and lipid/polymer dynamics [78], reinforced by reviews [48]. Aiello et al. [79] optimized CPMG for polymer low-MW species.

Acknowledgements

The author wishes to thank Slovenian Research Agency ARIS's research program P1-0060 and Prof. Janez Stepišnik.

Author details

Aleš Mohorič[1,2]

1 Faculty of Mathematics and Physics, University of Ljubljana, Ljubljana, Slovenia

2 Jozef Stefan Institute, Ljubljana, Slovenia

*Address all correspondence to: ales.mohoric@fmf.uni-lj.si

IntechOpen

References

[1] Bloch F. Nuclear induction. Physical Review. 1946;**70**(7–8):460-474. DOI: 10.1103/PhysRev.70.460

[2] Bloembergen N, Purcell EM, Pound RV. Relaxation effects in nuclear magnetic resonance absorption. Physical Review. 1948;**73**:679-712. DOI: 10.1103/PhysRev.73.679

[3] Carr HY, Purcell EM. Effects of diffusion on free precession in nuclear magnetic resonance experiments. Physical Review. 1954;**94**(3):630-638. DOI: 10.1103/PhysRev.94.630

[4] Meiboom S, Gill D. Modified spin-echo method for measuring nuclear relaxation times. Review of Scientific Instruments. 1958;**29**(8):688-691. DOI: 10.1063/1.1716296

[5] Loria JP, Rance M, Palmer AG. A relaxation-compensated CPMG sequence for characterization of chemical exchange by NMR. Journal of the American Chemical Society. 1999;**121**(10):2331-2332. DOI: 10.1021/ja983961a

[6] Redfield AG. On the theory of relaxation processes. IBM Journal of Research and Development. 1957;**1**(1): 19-31. DOI: 10.1147/rd.11.0019

[7] McConnell HM. Reaction rates by nuclear magnetic resonance. Journal of Chemical Physics. 1958;**28**(3):430-431. DOI: 10.1063/1.1744152

[8] Carver JP, Richards RE. A general two-site solution for the chemical exchange produced dependence of T2 upon the Carr-Purcell pulse separation. Journal of Magnetic Resonance. 1972;**6** (1):89-105. DOI: 10.1016/0022-2364(72) 90090-X

[9] Bain AD. Chemical exchange in NMR. Progress in Nuclear Magnetic Resonance

Spectroscopy. 2003;**43**(3-4):63-103. DOI: 10.1016/j.pnmrs.2003.08.001

[10] Forsén S, Hoffman R. Study of moderately rapid chemical exchange reactions by means of nuclear magnetic double resonance. The Journal of Chemical Physics. 1963;**39**(11):2892-2901. DOI: 10.1063/1.1734121

[11] Kay LE, Torchia DA, Bax A. Backbone dynamics of proteins studied by 15N inverse detected heteronuclear NMR spectroscopy. Biochemistry. 1989; **28**(23):8972-8979. DOI: 10.1021/ bi00449a003

[12] Tollinger M et al. Slow dynamics in folded and unfolded states of an SH3 domain. Journal of the American Chemical Society. 2001;**123**(46):57. 11341-11352. DOI: 10.1021/ja011300z

[13] Clore GM, Gronenborn AM. Two-, three-, and four-dimensional NMR methods for obtaining larger and more precise three-dimensional structures of proteins in solution. Annual Review of Biophysics. 1991;**20**:29-63. DOI: 10.1146/ annurev.bb.20.060191.000333

[14] Palmer AG Jr. Nuclear magnetic resonance methods for quantifying microsecond-to-millisecond motions in biological macromolecules. Methods in Enzymology. 2001;**339**: 204-238. DOI: 10.1016/s0076-6879(01) 39315-1

[15] Palmer AG. NMR characterization of the dynamics of biomacromolecules. Chemical Reviews. 2004;**104**(8):3623-3640. DOI: 10.1021/cr030413t

[16] Wangsness RK, Bloch F. The dynamical theory of nuclear induction. Physics Review. 1953;**89**:728-739. DOI: 10.1103/PhysRev.89.728

[17] Levitt MH. Spin Dynamics: Basics of Nuclear Magnetic Resonance. Chichester, England: John Wiley & Sons; 2008

[18] Ernst RR, Bodenhausen G, Wokaun A. Principles of Nuclear Magnetic Resonance in One and Two Dimensions. Oxford, England: Oxford University Press; 1987

[19] Mohorič A, Stepišnik J. Chemical exchange rate study by NMR CPMG method. Applied Magnetic Resonance. 2023;**54**:1411-1422. DOI: 10.1007/s00723-023-01621-z

[20] Mittermaier A, Kay LE. New tools provide new insights in NMR studies of protein dynamics. Science. 2006;**35**(5771):224-228. DOI: 10.1126/science.1124964

[21] Mulder F et al. Studying excited states of proteins by NMR spectroscopy. Nature Structural & Molecular Biology. 2001;**8**:932-935. DOI: 10.1038/nsb1101-932

[22] Grzesiek S, Bax A. Correlating backbone amide and side chain resonances in larger proteins by multiple relayed triple resonance NMR. Journal of the American Chemical Society. 1992;**114**(16):6291-6293. DOI: 10.1021/ja00042a003

[23] Palmer AG. Chemical exchange in biomacromolecules: Past, present, and future. Journal of Magnetic Resonance. 2014;**241**:3-17. DOI: 10.1016/j.jmr.2014.01.008

[24] Ashish A et al. Chapter 5 – Advanced NMR spectroscopy methods to study protein structure and dynamics. In: Saudagar P, Tripathi T, editors. Advanced Spectroscopic Methods to Study Biomolecular Structure and Dynamics. USA: Academic Press,

Elsevier; 2022. pp. 125-152. ISBN: 978-032-39-9127-8. DOI: 10.1016/C2021-0-01551-0

[25] Hansen DF et al. Probing chemical shifts of invisible states of proteins with relaxation dispersion NMR spectroscopy: How well can we do? Journal of the American Chemical Society. 2008;**130**(8):2667-2675. DOI: 10.1021/ja078337p

[26] Matviychuk Y et al. Time-domain signal modelling in multidimensional NMR experiments for estimation of relaxation parameters. Journal of Biomolecular NMR. 2019;**73**:93-104. DOI: 10.1007/s10858-018-00224-2

[27] Korzhnev DM et al. Probing slow dynamics in high molecular weight proteins by methyl-TROSY NMR spectroscopy: Application to a 723-residue enzyme. Journal of the American Chemical Society. 2004;**126**(12):3964-3973. DOI: 10.1021/ja039587i

[28] Loria JP et al. A TROSY CPMG sequence for characterizing chemical exchange in large proteins. Journal of Biomolecular NMR. 1999;**15**:151-155. DOI: 10.1023/A:1008355631073

[29] Vallurupalli P, Hansen DF, Kay LE. Structures of invisible, excited protein states by relaxation dispersion NMR spectroscopy. Proceedings of the National Academy of Sciences. 2008;**105**(33):11766-11771. DOI: 10.1073/pnas.0804221105

[30] Chao F, Byrd RA. Protein dynamics revealed by NMR relaxation methods. Emerging Topics in Life Sciences. 2018;**2**(1):93-105. DOI: 10.1042/etls20170139

[31] Palmer AG 3rd, Koss H. Chemical exchange. Methods in Enzymology. 2019;**615**:177-236. DOI: 10.1016/bs.mie.2018.09.028

[32] Farrow NA et al. Backbone dynamics of a free and phosphopeptide-complexed Src homology 2 domain studied by 15N NMR relaxation. Biochemistry. 1994;**33** (19):5984-6003. DOI: 10.1021/ bi00185a040

[33] Korzhnev DM, Religa TL, Banachewicz W, Fersht AR, Kay LE. A transient and low-populated protein-folding intermediate at atomic resolution. Science. 2010;**329**(5997): 1312-1316. DOI: 10.1126/science.1191723

[34] Korzhnev DM, Salvatella X, Vendruscolo M, Di Nardo AA, Davidson AR, Dobson CM, et al. Low-populated folding intermediates of Fyn SH3 characterized by relaxation dispersion NMR. Nature. 2004;**430**(6999):586-590. DOI: 10.1038/nature02655

[35] Vallurupalli P et al. Structures of invisible, excited protein states by relaxation dispersion NMR spectroscopy. Biophysics and Computational Biology. 2008;**105**(33):11766-11771. DOI: 10.1073/ pnas.0804221105

[36] Vallurupalli P et al. Studying "invisible" excited protein states in slow exchange with a major state conformation. Journal of the American Chemical Society. 2012;**134**(19):8148-8161. DOI: 10.1021/ja3001419

[37] Farrow NA et al. Spectral density-function mapping using N-15 relaxation data exclusively. Journal of Biomolecular NMR. 1995;**6**(2):153-162. DOI: 10.1007/ BF00211779

[38] Koss H et al. General expressions for Carr-Purcell-Meiboom-Gill relaxation dispersion for N-site chemical exchange. Biochemistry. 2018;**57**(31):4753-4763. DOI: 10.1021/acs.biochem.8b00370

[39] Koss H et al. Algebraic expressions for Carr-Purcell-Meiboom-Gill

relaxation dispersion for N-site chemical exchange. Journal of Magnetic Resonance. 2020;**321**:106846. DOI: 10.1016/j.jmr.2020.106846

[40] Massi F, Peng JW. Characterizing protein dynamics with NMR R 1ρ relaxation experiments. Methods in Molecular Biology. 2018;**1688**:205-221. DOI: 10.1007/978-1-4939-7386-6_10

[41] Clore GM et al. Elucidating transient macromolecular interactions using paramagnetic relaxation enhancement. Current Opinion in Structural Biology. 2007;**17**:603-616. DOI: 10.1016/j. sbi.2007.08.013

[42] Kay LE et al. Pure absorption gradient enhanced heteronuclear single quantum correlation spectroscopy with improved sensitivity. Journal of the American Chemical Society. 1992;**114** (26):10663-10665. DOI: 10.1021/ ja00052a088

[43] Hansen DF et al. Implications of using approximate Bloch–McConnell equations in NMR analyses of chemically exchanging systems. Journal of Magnetic Resonance. 2003;**163**(2):215-227. DOI: 10.1016/S1090-7807(03)00062-4

[44] Bolik-Coulon N et al. Less is more: A simple methyl-TROSY based pulse scheme offers improved sensitivity in applications to high molecular weight complexes. Journal of Magnetic Resonance. 2023;**346**:107326. DOI: 10.1016/j.jmr.2022.107326

[45] Loria JP et al. A relaxation-compensated Carr-Purcell-Meiboom-Gill sequence for characterizing chemical exchange by NMR spectroscopy. Journal of the American Chemical Society. 1999; **121**(10):2331-2332. DOI: 10.1021/ ja983961a

[46] Watt ED et al. The mechanism of rate-limiting motions in enzyme

function. Proceedings of the National Academy of Sciences of the United States of America. 2007;**104**(29):11981-11986. DOI: 10.1073/pnas.0702551104

[47] Mulder FAA. NMR spectroscopy charges into protein surface electrostatics. Proceedings of the National Academy of Sciences of the United States of America. 2021;**118**(30): e2110176118. DOI: 10.1073/ pnas.2110176118

[48] Palmer AG 3rd. A dynamic look backward and forward. Journal of Magnetic Resonance. 2016;**266**:73-80. DOI: 10.1016/j.jmr.2016.01.018

[49] Reddy JG, Griesinger C, et al. Simultaneous determination of fast and slow dynamics in molecules using extreme CPMG relaxation dispersion experiments. Journal of Biomolecular NMR. 2018;**70**(1):1-9. DOI: 10.1007/ s10858-017-0155-0

[50] Bostock MJ, Holland DJ, Nietlispach D. Improving resolution in multidimensional NMR using random quadrature detection with compressed sensing reconstruction. Journal of Biomolecular NMR. 2017;**68**:67-77. DOI: 10.1007/s10858-016-0062-9

[51] Smith AM et al. Quantitative analysis of thiolated ligand exchange on gold nanoparticles monitored by 1H NMR spectroscopy. Analytical Chemistry. 2015;**87**(5):2771-2778. DOI: 10.1021/ ac504081k

[52] Alaei Z et al. Solvent relaxation NMR as a tool to study particle dispersions in non-aqueous systems. Physchem. 2022;**2** (3):224-234. DOI: 10.3390/ physchem2030016

[53] Zhang R. Reversible cross-linking, microdomain structure, and heterogeneous dynamics in thermally

reversible cross-linked polyurethane as revealed by solid-state NMR. The Journal of Physical Chemistry B. 2014;**118**(4): 1126-1137. DOI: 10.1021/JP409893F

[54] Banyai I. Dynamic NMR for coordination chemistry. New Journal of Chemistry. 2018;**42**:7569-7581. DOI: 10.1039/C8NJ00233A

[55] Deev SL et al. 15N labeling and analysis of 13C–15N and 1H–15N couplings in studies of the structures and chemical transformations of nitrogen heterocycles. RSC Advances. 2019;**9**: 26856-26879. DOI: 10.1039/ C9RA04825A

[56] Wang C et al. CPMG sequences with enhanced sensitivity to chemical exchange. Journal of Biomolecular NMR. 2001;**21**:361-366. DOI: 10.1023/A: 1013328206498

[57] Anderson JL et al. Characterizing ionic liquids on the basis of multiple solvation interactions. Journal of the American Chemical Society. 2002;**124** (47):14247-14254. DOI: 10.1021/ ja028156h

[58] Li Y et al. The synthesis and characterization of polyacrolein through radical polymerization. Macromolecular Reaction Engineering. 2021;**15**(1): 2000046. DOI: 10.1002/ mren.202000046

[59] Agarwal N et al. Chapter 3 - Characterization of nanomaterials using nuclear magnetic resonance spectroscopy. In: Bhagyaraj SM et al., editor. Micro and Nano Technologies, Characterization of Nanomaterials. Woodhead Publishing; 2018. pp. 61-102. ISBN 9780081019733. DOI: 10.1016/ B978-0-08-101973-3.00003-1

[60] Holzgrabe U et al. NMR spectroscopy in pharmacy. Journal of

Pharmaceutical and Biomedical Analysis. 1998;**17**(4-5):557-616. DOI: 10.1016/S0731-7085(97)00276-8

[61] Tromp RR, Cerioni LMC. Multiphase flow regime characterization and liquid flow measurement using low-field magnetic resonance imaging. Molecules. 2021;**26**:5702. DOI: 10.3390/molecules26113349

[62] Cônsolo NRB et al. Characterization of chicken muscle disorders through metabolomics, pathway analysis, and water relaxometry: A pilot study. Poultry Science. 2020;**99**:6247-6257. DOI: 10.1016/j.psj.2020.06.066

[63] de Andrade FD et al. Qualitative analysis by online nuclear magnetic resonance using Carr–Purcell–Meiboom–Gill sequence with low refocusing flip angles. Talanta. 2011;**84**(1):84-88. DOI: 10.1016/j.talanta.2010.12.033

[64] Jarenwattananon NN, Bouchard LS. Breakdown of Carr-Purcell Meiboom-Gill spin echoes in inhomogeneous fields. Journal of Chemical Physics. 2018;**149**:084304, 1–4. DOI: 10.1063/1.5043495

[65] Hürlimann MD et al. Spin dynamics of the Carr-Purcell-Meiboom-Gill sequence in time-dependent magnetic fields. Physical Review Applied. 2019;**12**(4):044061. DOI: 10.1103/PhysRevApplied.12.044061

[66] Cistola DP, Robinson MD. Compact NMR relaxometry of human blood andb lood components. Trends in Analytical Chemistry. 2016;**83**(A):53-64. DOI: 10.1016/j.trac.2016.04.020

[67] Shomaji S et al. Detecting dye-contaminated vegetables using low-field NMR relaxometry. Foods. 2021;**10**(9):2232. DOI: 10.3390/foods10092232

[68] Eberhardt TL et al. Analysis of ethanol-soluble extractives in southern pine wood by low-field proton NMR. Journal of Wood Chemistry and Technology. 2007;**27**(1):35-47. DOI: 10.1080/02773810701285622

[69] Gottwald A et al. Diffusion, relaxation, and chemical exchange in casein gels: A nuclear magnetic resonance study. Journal of Chemical Physics. 2005;**122**(3):34506. DOI: 10.1063/1.1825383

[70] Casieri C et al. Pore-size evaluation by single-sided nuclear magnetic resonance measurements: Compensation of water self-diffusion effect on transverse relaxation. Journal of Applied Physics. 2005;**97**:043901. DOI: 10.1063/1.1833572

[71] Al-Mahrooqi SH et al. An investigation of the effect of wettability on NMR characteristics of sandstone rock and fluid systems. Journal of Petroleum Science and Engineering. 2003;**39**(3–4):389-398. DOI: 10.1016/S0920-4105(03)00077-9

[72] Lefort R et al. Optimization of data acquisition and processing in Carr–Purcell–Meiboom–Gill multiple quantum magic angle spinning nuclear magnetic resonance. Journal of Chemical Physics. 2002;**116**:2493-2501. DOI: 10.1063/1.1433000

[73] Utsuzawa S et al. Ringing cancellation in Carr-Purcell-Meiboom-Gill-type sequences. Magnetic Resonance Letters. 2022;**2**(4):233-242. DOI: 10.1016/j.mrl.2022.03.002

[74] Bai R et al. Fast, accurate 2D-MR relaxation exchange spectroscopy (REXSY): Beyond compressed sensing. Journal of Chemical Physics. 2016;**145**(15):154202. DOI: 10.1063/1.4964144

[75] ISO 16756:2024. Milk and Milk Products—Guidance for the Application of Carr-Purcell-Meiboom-Gill (CPMG) Pulsed Time-Domain Nuclear Magnetic Resonance (TD-NMR) Spectroscopy for Fat Determination. Geneva, Switzerland: ISO Standards; 2024

[76] Dixon AM, Larive CK. NMR spectroscopy with spectral editing for the analysis of complex mixtures. Applied Spectroscopy. 1999;**53**(11): 426A-440A. DOI: 10.1366/0003702991945704

[77] Hayes MHB et al. Chapter two - Humin: Its composition and importance in soil organic matter. In: Sparks DL, editor. Advances in Agronomy. 2017; **143**:47-138. DOI: 10.1016/bs.agron.2017.01.001

[78] Narayanan C et al. Applications of NMR and computational methodologies to study protein dynamics. Archives of Biochemistry and Biophysics. 2017;**628**: 71-80. DOI: 10.1016/j.abb.2017.05.002

[79] Aiello F et al. A multivariate approach to investigate the NMR CPMG pulse sequence for analysing low MW species in polymers. Magnetic Resonance in Chemistry. 2021;**59**:172-186. DOI: 10.1002/mrc.5100

Chapter 5

In Vivo, Magnetic Resonance Spectroscopy (1-HRM)-Derived Autism-Related Neurotransmitter Variations

Carmen Jiménez-Espinoza, Francisco Marcano Serrano and José González-Mora

Abstract

Autism spectrum disorder (ASD) is a severe developmental syndrome of unclear etiology, arising largely as a disorder of neural systems. One of the most studied causes is the increased excitation-inhibition in sensory and social systems, which may explain certain phenotypic expressions in ASD related to different neurotransmitters and their imbalances (NT). This study aims to assess neurotransmitter levels in the anterior cingulate (ACC) and posterior cingulate (PCC) cortices in subjects with ASD using *in vivo* proton magnetic resonance spectroscopy (1H-MRS), considered a robust tool. The finding of an imbalance in the neurotransmitters myoinositol (mI), choline (Cho), N-acetyl aspartil glutamate (NAA + NAAG), and glutamate (Glu) in cingulated cortices sparked our interest as an important cause of their multiple etiology, providing strong empirical support for increased arousal in ASD. This metabolic imbalance between ACC and PCC was correlated using the autism quotient (AQ) test, suggesting potential therapeutic interventions.

Keywords: autism spectrum disorders, excitotoxicity, cingulated cortices, spectroscopy resonance magnetic, social skills, attention switching/tolerance to change, attention to detail, imagination, communication

1. Introduction

Autism is related to the triad of social, communicative, imaginative, and behavioral limitations described by Wing and Gould, 1979 [1]. However, recent evidence shows that autism presents a great phenotypic variation that gives it the identity of a multifactorial disease. This has led to the latest revision of the Diagnostic and Statistical Manual of Mental Disorders DSM-V [2], where autism has been renamed autism spectrum disorder (ASD) as a single category. Although the new rename of autism has increased the diagnostic accuracy with more accuracy, using more accurate procedures and instruments, that has improved the knowledge and training of professionals, the real increase in the incidence of this disorder may be due to the increased global population. In this sense, the development of new diagnostic methods to quantify and study the evolution of the spectrum of autism in each patient could help us better understand the neurometabolic processes involved more reliably. From the physiological point of view, brain metabolism comprises a wide variety of molecules, including peptides, neurotransmitters, enzymes, all interacting with water. These molecules can maintain the physiological balance necessary for the healthy functioning of the brain due to their functions and activity. Furthermore, the brain is one of the organs with the highest energy demand, consuming 20% of the oxygen and 25% of the glucose ingested by the body [3, 4]. Considering that, all this neurometabolic exchange must be continued, fast, and effective, precisely because the brain has few energy reserves and only receives them through the cerebral vascular system, which involves a correct supply of cerebral circulation through the circle of Willis.[1]

The Proton Magnetic Resonance Spectroscopy (1H-MRS) is a technique available in radiological practice that provides a biochemical, metabolic, and functional assessment of tissues considering that all the mechanisms and pathways that accompany synapses in the central nervous system (CNS) occur in a very short time scale (milliseconds). It is used principally in the evaluation of brain tumors and now has spread to the valuation of other diseases such as metabolic disorders, studies of dementia, vascular disorders, and assessment of some psychiatric disorders, and more recently in heart disease, liver, breast, and prostate assessment. Despite the consideration of autism as a multifactorial disease and the wide application of spectroscopy in some psychiatric disorders, there is still no consensus that allows the use of this tool for the assessment of ASD. Recent studies have indicated the existence of some defects in the synthesis, release, or degradation of some neurotransmitters involved in the pathogenesis of many neurological, muscular, and psychiatric diseases [5], directing our objective to the study, using 1H-MRS, of possible variations of NT in the brain tissue of people with ASD.

2. Neurotransmitter's characteristics and functions

A group of biomolecules in the brain can be detected using proton magnetic resonance spectroscopy. We are referring to neurotransmitters, defined as endogenous sub-signals that act as chemical messengers that transmit signals and are normally released by neurons into the synaptic space, where they exert their function on other neurons or other target cells through a synapse. In summary, at least 100 neurotransmitters (NT) have been described already, and many others have yet to be discovered. Neurotransmitters can transmit one of three possible actions: excitatory, inhibitory, and modulatory messages, depending on the NT function. The better-known categories and neurotransmitter examples and their functions are included in **Table 1**. Some are grouped into types according to their chemical nature and are used as markers in 1H-MRS.

Neurotransmitters are generally classified into two main categories based on their size: (a) small molecule neurotransmitters and (b) neuropeptides, related to their overall excitatory or inhibitory activity. Another group of biomolecules involved in

---

[1] Circle of Willis: Is a junction of several important arteries at the bottom part of the brain. It helps blood flow from both the front and back sections of the brain. https://www.medicalnewstoday.com/articles/circle-of-willis

Metabolite	Abbreviation	Cellular location	Function	Decrease	Increase
N-Acetyl-aspartate (6,7,8,9)	NAA	Different types of neurons, oligodendrocytes.	Neuronal marker. Amino acid precursor of NAAG together with glutamate. Its concentration varies in different types of neurons.	Loss of neurons or axons. Dementia, epilepsy, multiple sclerosis, ischemia, tumors. Manic states.	Canavan disease.
N-acetyl-aspartyl glutamate (8,9)	NAAG	Neurons, microglia, astrocyte, and oligodendrocyte.	Synapse-modulating peptide. Responsible, along with glial cells, for synaptic plasticity.	Damage to neurons in the motor system.	Neuropro-tection by reduces excess of glutamate.
Glutamate (10)	Glu	Presynaptic vesicles. Neurons, glia, and non-synaptic activities (cellular metabolism)	Excitatory-inhibitory neurotransmitter of the synapse. Glial marker. It shares with glutamine an intercellular cycle between neurons and glia.	Concentrate trouble mental exhaustion. Insomnia. Low energy.	Extra or intra cellular glial edema, tissue damage.
Glutamine (10)	Gln	Astrocytes	Glial marker. Acts as an inhibitory neurotransmitter at the brain level. Precursor glutathione.	Decreased γ-aminobutyrate and the enzyme glutamine synthetase.	Hyperactivity. Toxicity by cumulative ammonia.
Creatine (11,12,13)	Cr	Cytosol of muscles and neurons	Homeostasis of cellular bioenergetics (indicator of ATP formation).	Hyper-metabolic states. (Brain tumor)	Hypo-metabolism states
Choline (11,12,13)	Cho	Membrane of nerve and muscle cells.	Precursor of the neurotransmitter Acetylcholine. Muscular control and memory. Synthesis and destruction of the cell membrane (cell- turnover)	Deficient exchange of cell membranes. States of mania and depression.	Myelin- destruction, tumor proliferation gliomas, meningiomas
Myoinositol (11,12,13)	mI	Glia, Astrocytes	Glial marker. Cell signal transduction. Useful in the assessment of brain maturation.	Non-glial tumors. Manic states	Gliosis and reactive astrocytosis Alzheimer's
Glutathione 14,15,16,17	GSH	Cytoplasm of glial cells and neurons (higher concentration in neurons than in glia). Extracellular space	Cellular toxicity marker. Neurohormone and modulators. Cellular and mitochondrial protectors from oxidative damage and peroxidation. Cellular homeostasis.	coronary and autoimmune diseases, cancer, dysfunction mitochondrial, muscle weakness and fatigue.	High oxidative stress, immune deficiency, detoxification xenobiotics

Table 1.
A summary of the main metabolites detectable in the human brain, their functions, and the consequence of their imbalance is referenced in the following works [6–17].

brain metabolism are the proteins that constitute a group of "receptors" and "transporters" necessary for the metabolism of neurotransmitters. Receptors are proteins that allow certain neurotransmitters to interact with metabolic processes in the cell, or in the nerve ending that captures a stimulus and transmits it to produce a response, and are divided into two groups: (a) ligand-opened ion channels (Ionotropic) and (b) receptors coupled to G proteins (Metabotropic). Their effect can be excitatory if they tend to depolarize the membrane or inhibitory if they repolarize it, where after acting they are degraded or recaptured by the presynaptic cell quickly [18].

Likewise, neurotransmitter transporters are classified into two groups: (a) the reuptake transporter, located in presynaptic neurons, which pumps NT from the extracellular space into the cell; and (b) the transporter located in the vesicle membranes and concentrates NT in them for later exocytosis. Furthermore, another group of transporters, which have mostly been cloned and are grouped into two large families based on their molecular characteristics and electrogenic properties, called axolemmal is known. Those dependent on Na+/Cl- which include transporters for GABA, norepinephrine, dopamine, serotonin, choline, proline, betaine, glycine, and taurine, and those dependent on Na+/K+ include transporters for glutamate as well as alanine, serine, and cysteine [19].

Also, existing other molecules can be released from the same axon terminals as neurotransmitters, and are known as neuromodulators [20] and they can increase, prolong, inhibit, or limit the effect of the main neurotransmitter on the postsynaptic membrane, acting through a system of second messengers [21]. Among the endogenous neuromodulators also known as neuropeptides of the central nervous system, oxytocin acts by modulating social behaviors and its decrease has recently been linked to the appearance of autism [22].

Functionally, inhibitory properties of neurotransmitters are characterized by preventing or blocking chemical messages, decreasing the stimulation of nerve cells in the brain. For example, glutamate (Glu) acts as an "on switch" in neuronal pathways and requires the neurotransmitter GABA as an "off switch" providing the necessary balance between these two neurotransmitters for the correct functioning of the CNS, essentially in the regulation of cognition, learning, memory and emotional behaviors [23–25]. Thus, the imbalance between Glu-excitation and GABA-inhibition leads to hyperactivation of the CNS that has been linked to ASD symptoms.

3. Materials and methods

Neurotransmitter metabolism in cingulated cortices has aroused great interest in recent research on autism spectrum disorders (ASD). In this chapter, we present a study carried out in the cingulate cortices of autistic adults, one of the areas that make up the limbic system, because it is related to memory, attention, and behavior in humans, qualities that are affected in people with ASD. In both regions of the cingulate cortices, anterior (ACC) and posterior (PCC), we applied the proton magnetic resonance spectroscopy (1H-MRS) technique, as a powerful tool that provides biochemical, functional, and metabolic information that complements obtained from conventional imaging studies. Here, changes in the concentration levels of different brain metabolites, as markers of neuronal density, glial density, cell membrane processes, and energy metabolism, are evaluated between the anterior and posterior cingulate cortex of subjects with ASD.

3.1 Sample size and strength

This study was carried out under ethical standards, and the latest Declaration of Helsinki (revised in 2000) was approved by the University of La Laguna. This study aims to detect the levels of neurotransmitters present in brain-cingulated cortices in patients with ASD. For this purpose, an adult population was assessed with 1H-MRS between 17 and 23 (mean, 20.8 ± 1.8) years. The sample of subjects with ASD (n = 19) and subjects with typical development (TD) (n = 41) were chosen to make the comparisons. Previously, a small number of adults who met the criteria for ASD (n = 5) and TD (n = 1) were excluded from the analyses because they had epilepsy.

3.2 Sample recruitment, sample characteristics, and adherence to the study

The adults were selected and recruited by the respective psychiatry and psychology service of the Hospital Universitario de Canarias (HUC), Tenerife, Spain. They were diagnosed with ASD in childhood and during their growth by applying the tests used to evaluate and measure early childhood development [26–31] without any variation in diagnosis during growth according to clinical history. The Autism Diagnostic Interview-Revised (ADI-R), or the Autism Diagnostic Observation Schedule (ADOS), is the standardized diagnostic test for assessing autism spectrum disorder which, not applied independently by us in this study, considers that there was a high degree of concordance between clinical and research diagnoses [32], thus avoiding overdiagnosis. Despite this lack of diagnosis can be seen as a limitation of the methodology, we considered applying the Spanish version of the AQ (Autism-Spectrum Quotient) test (Baron-Cohen S, 2005) [33] to all participants for a significant differentiation between the groups in the study, namely, subjects with autism group (ASD) and subjects with typical development (TD) used as control (see **Table 2**).

3.3 1H-MRS acquisition data

All patients underwent MRI and 1H-MRS on a Signa HD 3 T MRI scanner (GE Healthcare, Waukesha, WI, USA). Single-voxel acquisition used a spin-echo sequence recorded within the following parameters: echo time (TE) = 23 ms, repetition time (TR) = 1070 ms, 2 NEX, flip angle = 90°, 256 acquisitions using the point-resolved spectroscopy (PRESS) technique. During the acquisitions, voxels (20 x 20 x 20 mm^3) were placed in ACC and PCC (see **Figure 1**), by the same experienced neuroradiologist blinded to the clinical data. We used TE-23 ms because myo-inositol is known

Characteristics	ASD (*n* = 19)	TD (*n* = 41)	*p* Value
Gender (male/female)	16/3	16/25	*p* = 0.016
Age(years)	20.58 (0.71)	23.19 (0.71)	*p* = 0.049
Epilepsy	5		*p* < 0.0015
Gastrointestinal disorders	17		*p* < 0.0001
Muscular hypotonia	14		*p* < 0.0001
Autism Quotient (0–50) points	33.84 (6.36)	11.67 (7.07)	*p* < 0.0001

Table 2.
Demographic and psychometric summary of participants.

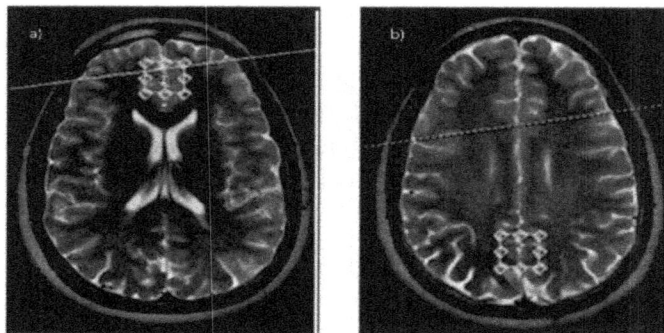

Figure 1.
Locations of the volume studied in the (a) anterior and (b) posterior cingulated cortices. The single voxel acquisition used a spin-echo sequence recorded within the following parameters: echo time (TE) =23 ms, repetition time (TR) =1070 ms, 2 NEX, flip angle = 90°, 256 acquisitions with point-resolved spectroscopy (PRESS) technique. While we obtained acquisitions, the same experienced neuroradiologist who was blind to the clinical data, placed voxels (20 x 20 x 20 mm³) at ACC and PCC.

to be easily detected in the short TE 1H-MRS spectra in the brain due to its high concentration of 4–8 mM [34]. Each voxel was carefully positioned in the study area in each subject to exclude signal contamination from the skull and subcutaneous fat. Furthermore, the morphological examination excluded other pathologies, such as congenital anomalies, cerebral palsy lesions, tumors, and hydrocephalus in this study.

3.4 Post-processing of the spectra

The data acquired by the magnetic resonance scanner was sent to the General Electric workstation (GE Workstation) in our image processing laboratory. There, the LC Model program [35] was installed on a Linux workstation dedicated to the automatic quantification of the spectra obtained. Also, the LC Mgui graphical interface, available to the user, was used to run the LC Model program more comfortably. Thus, the results of the processing with LC Model are obtained in a two-page printable file (postScript), for each voxel studied (see **Figure 2**) of results, where the concentration of each metabolite is summarized with its respective Cramér-Rao limit[2] [36]. In addition, before looking at the concentrations of the metabolites, it is important to first look at the standard deviation (expressed as a percentage, % SD) in the input column, since these have been estimated according to the lower limits of Cramér-Rao concentrations, and although they are only the lower limits, they are still the most useful reliability indicators. The reliability indicator represented by % SD $\leq$ 20 is subjectively considered as a very approximate criterion for acceptable reliability estimates.

The statistical package used to make the graphical representations and statistical analysis in this thesis was the GraphPad Prisma v 10.4.0 program (GraphPad Software, inc. La Jolla, San Diego. USA).

3.5 Neurometabolite post-processing and quantification

1H-MRS datasets were collected using the SAGE and LCMODEL 6.3 software platforms considering that the calculation of absolute concentrations (e.g., in molar

---

[2] Cramér-Rao lower bounds. These are approximate maximum likelihood estimates and their uncertainties. Provencher SW. 1993

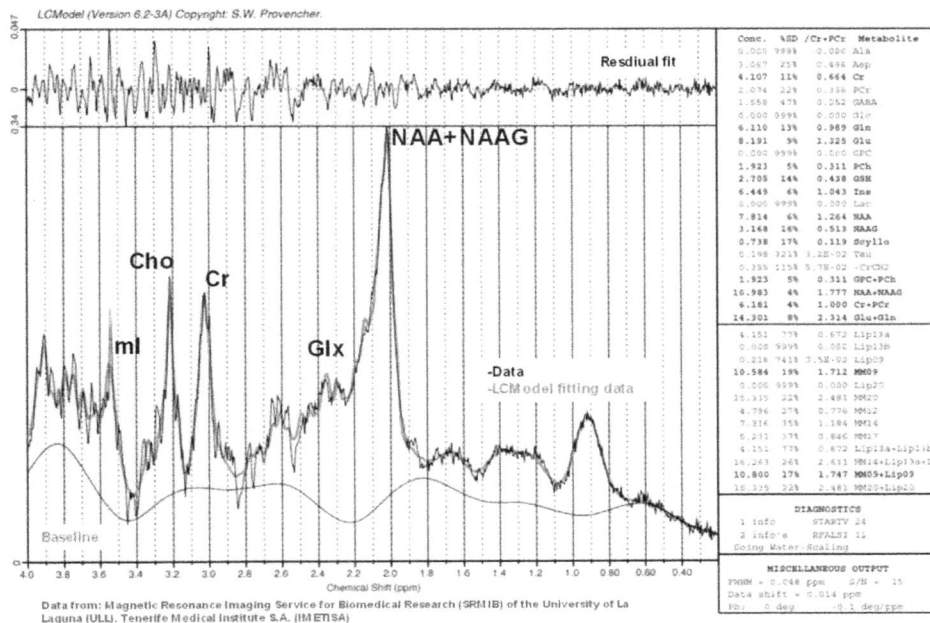

Figure 2.
Post-processed spectrum in a single voxel, obtained by LCMODEL. MR spectrum acquired showing the prominent resonances of the brain metabolites' N-acetyl aspartate+N-acetyl aspartil glutamate (NAA + NAAG), Glutamate+Glutamine (Glx), creatine (Cr), Choline (Cho), myoinositol (mI). The black line, looks like noise one is the actually data, and the red line is the LCModel fitting data. The noisy trace at the top (black) is the residual fit. The numbers on the right quantification results, including error estimates (%SD) in Cramér-Rao lower bounds (CRLBs).

units) involves correction for many factors, including the tissue composition of the voxel (relative amounts of cerebrospinal fluid and gray and white matter), the T1 and T2 relaxation times of the metabolites in the subject under study, the location of the voxels and their relationship to the electromagnetic properties of the coil, and any temporal variations in the scanner. However, the assessment of neurometabolite proportion with creatine is generally the fastest and most widely used analysis method for clinical 1H-MRS. Furthermore, ratios account for all non-metabolite-specific differences, therefore can be reasonably compared across all participants scanned at the same institution with the same protocol.

4. Results and discussion

4.1 Neurometabolite variability between anterior (ACC) and posterior (PCC) cingulated cortices in subjects with and without ASD. Proton spectroscopy (1H-MRS)

The resonance signals in the spectra of the detected neurometabolites are fitted by the curvature under the area of each peak shown within the voxel (see **Figures 1** and **2**). The absolute concentrations [mM] of the neurotransmitters present (see **Table 3**) were calculated using creatine as an internal reference because it is the most stable metabolite under the applied magnetic field [37]. Spectroscopy

Metabolites absolute concentration	ACC	PCC	P
ASD	[mM]	[mM]	
NAA + NAAG	10.29	10.87	n.s
NAA	9.70	10.53	n.s
Glx	16.84	13.92	n.s
Glu	12.60	10.29	*p = 0.04
Cho	2.15	1.56	*p = 0.0008
Cr	7.20	6.76	n.s
mI	5.24	5.03	n.s
GSH	3.08	3.43	n.s
TD			
NAA + NAAG	10.63	10.73	n.s
NAA	9.95	10.40	n.s
Glx	9.55	8.40	n.s
Glu	9.98	9.83	n.s
Cho	1.53	1.07	*p = 0.02
Cr	7.43	6.73	n.s
mI	5.55	4.94	*p = 0.03
GSH	3.75	3.31	n.s

Case-control analysis of metabolites values according to cerebral regions in ASD vs. TD, for N-acetyl aspartate (NAA), N-acetyl aspartate+N-acetyl aspartil glutamate (NAA + NAAG), Glutamate+Glutamine (Glx), Glutamate (Glu), creatine (Cr), Choline (Cho), myoinositol (mI), and Glutathione (GSH). Absolute concentration group means. Regions = Anterior cingulate cortex (ACC) and posterior cingulate cortex (PCC). P < 0.05 significantly differences. No significant differences = n.s.

Table 3.
Absolute concentrations [mM] of the main metabolites detected in ACC and PCC, within each group ASD and TD.

studies were performed on subjects in the ASD and TD groups, calculating the averages of the absolute concentrations of the metabolites detected in the voxel positioned in ACC and PCC. The resulting neurometabolic report is shown in **Table 4**.

When observing the pattern of neurometabolites obtained in the TD group and the variation in absolute concentrations in ACC and PCC, significant differences were observed in the metabolites myoinositol (p = 0.03) and choline (p = 0.02) between ACC and PCC (see **Table 3**). This describes the metabolic needs in each, related to the function of each area. On the one hand, myoinositol is considered a glial marker responsible for cellular signal transduction and useful for assessing brain maturation, and on the other hand, choline is responsible for maintaining the structural integrity of membranes and acts as a modulator of cholinergic neurotransmission. Furthermore, the function of choline is critical during fetal development, influencing the risk of neural tube defects and lifelong memory function [38]. We then considered this metabolic pattern present in healthy subjects with typical development as a control for comparing them with the results obtained in the group with ASD.

Contrary to the result of the TD group, the neurometabolite pattern obtained in the ASD group showed a significant increase in absolute glutamate (Glu) concentration (12.85; p = 0.04) in the ACC, indicating an imbalance of glutamatergic

Metabolite [mM] ± SD	AQ1 Score = (0–10)	AQ2 Score = (11–22)	AQ3 Score = (23–31)	AQ4 Score = (32–50)
ACC				
Cr + PCr	7.48 ± 0.81	7.42 ± 1.86	6.53 ± 0.98	7.72 ± 1.77
mI	6.12 ± 0.81	5.47 ± 1.42	7.29 ± 3.61	4.83 ± 1.93
GPC + PCho	1.83 ± 0.49	2.26 ± 0.59	1.63 ± 0.70	2.18 ± 0.63
NAA + NAAG	10.41 ± 0.79	10.66 ± 2.73	8.46 ± 0.49	10.56 ± 2.57
Glu + Gln	13.74 ± 4.06	15.67 ± 4.13	17.34 ± 1.43	16.65 ± 7.24
GSH	3.51 ± 0.87	3.96 ± 1.32	2.21 ± 0.84	3.36 ± 0.60
Glu	8.04 ± 1.62	10.25 ± 2.70	13.87 ± 2.13	13.14 ± 4.14
NAA	9.96 ± 1.27	9.95 ± 2.38	8.41 ± 0.38	9.71 ± 1.78
PCC				
Cr + PCr	7.05 ± 0.45	6.69 ± 1.38	6.21 ± 0.69	6.68 ± 0.51
mI	5.99 ± 1.79	4.80 ± 0.71	5.51 ± 0.99	4.91 ± 0.83
GPC + PCho	1.63 ± 0.39	1.66 ± 0.41	1.29 ± 0.36	1.48 ± 0.32
NAA + NAAG	10.28 ± 2.60	10.79 ± 2.08	9.60 ± 0.95	10.98 ± 1.24
Glu + Gln	13.37 ± 2.11	13.76 ± 0.83	10.39 ± 0.37	14.14 ± 5.45
GSH	3.08 ± 1.07	3.36 ± 0.91	2.50 ± 1.11	3.44 ± 0.94
Glu	10.51 ± 2.48	9.74 ± 2.15	8.94 ± 2.66	10.10 ± 2.20
NAA	9.89 ± 2.16	10.47 ± 2.04	8.98 ± 1.21	10.57 ± 1.56

Table 4.
Metabolite concentrations are grouped by AQ score (AQ1, AQ2, AQ3, and AQ4) in ACC and PCC regions. Mean ± SD.

metabolism (see **Table 3**), and confirming a large body of research results proposing this glutamate imbalance as one of the many etiologies of ASD. In this regard, a hyperglutamatergic state has been previously reported in ASD subjects, in late childhood and adolescence, specifically in the anterior cingulate cortex [39] further supporting our results. It is important to consider that glutamatergic gliotransmission is controlled by gliotransmitters released by oligodendrocytes and microglia [40–42], being the main responsible for excitotoxicity in the CNS; hence, it is related to a deregulation of the excitatory/inhibitory response in the brain as proposed by the new models of neurodevelopmental deficits in ASD. Glutamate, together with aspartate, constitutes the main transmitter of the dominant neurons of the cortex (called pyramidal cells), where developmental changes occur in the transitional stages of the visual cortex, affecting the nature of spatial and temporal integration in neurons that contribute to the maturation of visual functions [43]. Recently, the effects of excess glutamate in the anterior cingulate cortex have been directly correlated with attention shifting/shift tolerance in subjects with ASD [44].

However, although glutamate contributes to the development of the nervous system and its synaptic plasticity, important physiological failure within the human brain is excitotoxicity, attributed precisely to an accumulation of glutamate that contributes to a large number of disorders such as ischemia, epilepsy, AIDS-associated dementia, and the appearance of neurodegenerative pathologies such as Alzheimer's disease (AD) [45], Huntington's disease [46], Parkinson's disease [47], among others [48] highlighting the importance of adequate glutamatergic metabolism at the brain level. It should be noted that glutamate is considered the main and most important excitatory neurotransmitter within the CNS, which shares with glutamine an intercellular cycle between neurons and glia [49, 50].

Highlighting, the choline (Cho) absolute concentration differences between ACC and PCC are very significant ($p = 0.0008$) in comparison with the TD group suggesting hypermetabolism in membrane turnover, also associated with myelin destruction [51, 52], related to a metabolic imbalance between these two brain areas (see **Figure 3**). Some characterized choline functions include sustained attention, sleep/wake regulation, and learning and memory within the CNS [51], which are related to principal ASD symptoms. Cholinergic innervation of the human cerebral cortex arises in the basal forebrain and reaches its highest density in the limbic system components where they branch off into the limbic cortex formed by the cingulate

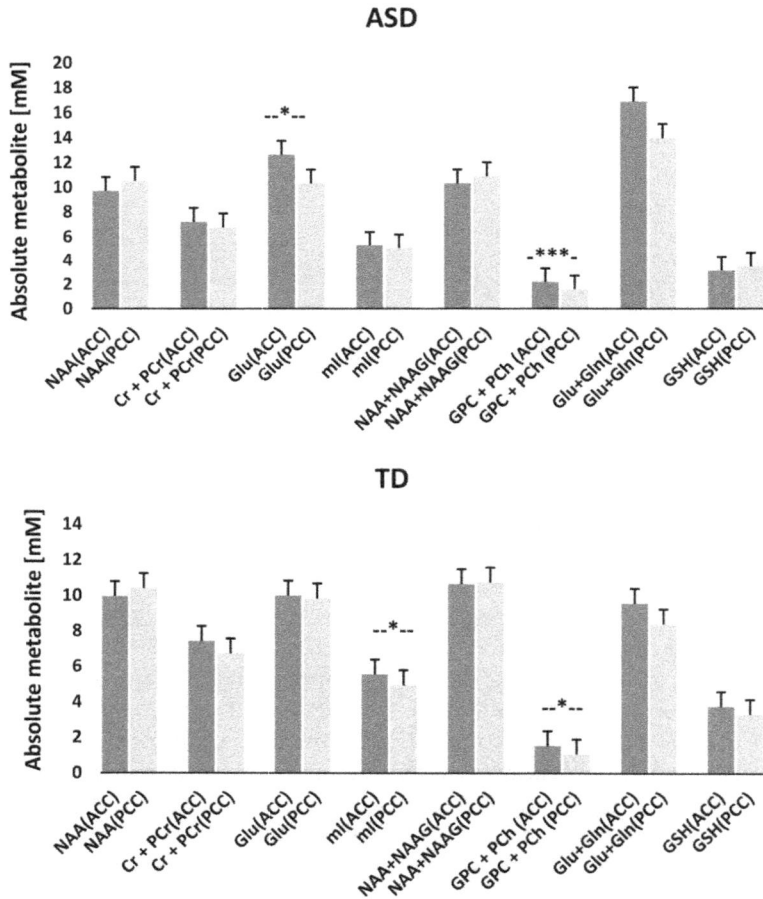

Figure 3.
Metabolic pattern in cingulated cortices ACC and PCC by ASD and TD groups. Mean absolute concentrations of the metabolites N-acetyl aspartate (NAA), N-acetyl aspartate+N-acetyl aspartyl glutamate (NAA + NAAG), Glutamate+Glutamine (Glx), Glutamate (Glu), creatine (Cr), Choline (Cho), myoinositol (mI), and Glutathione (GSH) in each ASD group (n = 19) and TD (n = 41). Comparison between the regions of the anterior and posterior cingulate cortex. Mann-Whitney test of Absolute NT. *p < 0.05; **p < 0.001; ***p < 0.0001.

gyrus. The behavioral affiliations of cortical cholinergic pathways span the domains of attention and memory [52], both of which are deficient in ASD [53].

Another important result in the ASD group was the significantly reduced concentration of myoinositol (mI) in ACC compared to the TD group. Neurochemical changes of mI in the anterior cerebral cortex have previously been associated with circadian disruption and depression in young adults [54] in agreement with our results. In more severe cases within ASD, the subjects are found to be disrupted to sleep during the night, increasing stress and inflammation in the brain. Myoinositol is found in glial cells and astrocytes controlling gliosis and reactive astrocytosis [55, 56]. Both cells interact with synapses that play important physiological roles in learning and memory [57] so a deficiency of mI in the ACC of subjects with ASD would be the cause of a lower performance in both functions [58].

Given the evidence, knowing the neurochemical substrate of both cingulate cortices, specifically, the PCC, which has a cerebral blood flow and metabolic rate of approximately 40% above the average of the entire brain [59], places it in the focus of our research, for understanding the relationship of activity that exists between both nodes (see **Figure 4**).

Taking into account ACC is the region that receives projections primarily from the amygdala and together with the dorsolateral prefrontal cortex, it intervenes in the regulation of behavior and plays an important role in shifting attention during the operation of working memory. Consequently, lesions in the anterior cingulate interfere with selective attention, monitoring of competing responses, and self-initiation of behavior, all skills that are deficient in subjects with ASD. In contrast to the ACC, the posterior cingulate cortex (PCC) receives most of the projections from the hippocampus, forming the emotional-social component of the memory system [60]. In this way, we present these two brain regions as substantially important for the investigation of ASD etiology.

On the other hand, ASD has been linked to a high vulnerability to oxidative stress and low capacity of methylation reaction with ASD symptoms in children [61, 62]. However, recent studies reveal that oxidative stress is related to a decrease in GSH concentration seen in peripheral blood mononuclear cells, plasma, lymphoblastoid cell lines, brain tissue, and mitochondria in individuals with ASD [63, 64] and that support

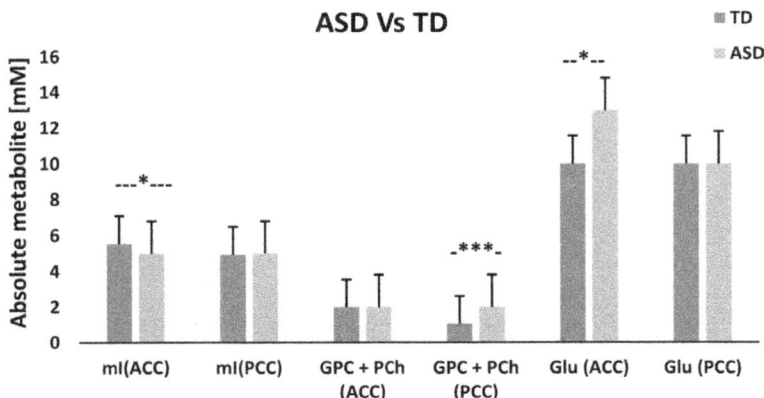

Figure 4.
Neurochemistry metabolite variation of absolute concentrations of Glutamate (Glu), Choline (Cho), and myoinositol (mI) in ASD group (n = 19) compared to TD (n = 41). Comparison between the regions of the anterior and posterior cingulate cortex. Mann-Whitney test of Absolute NT. *p < 0.05; **p < 0.001; ***p < 0.0001.

the results in this study, identifying glutathione as a biomarker of oxidative stress caused by an alteration in its metabolism, key in the etiology of ASD [14, 16, 64–66].

4.2 Neuroplasticity[3] in ACC and PCC is associated with the intensity of autism features according to the AQ-test score and the corresponding neurometabolic pattern

Our results suggest the usefulness of the AQ-Test in characterizing subjects within the broad spectrum of autism. The subjects were then grouped according to the range of the AQ-Test autism coefficient described by Baron-Cohen [67] which provides a broader view of the variations within the autism spectrum. Within the broad social spectrum in which autism traits are reflected, and their distribution within the population as a whole, the autism quotient (AQ)[4] represents the position an individual occupies within that broad spectrum or social continuum. Likewise, the author of the AQ Test attributes two essential uses to it:

1. In clinical practice as a screening tool, allowing physicians or psychologists to identify individuals potentially affected by ASD for subsequent referral for a more precise diagnosis; and

2. In research, in comparative studies, to separate "affected" from "non-affected" individuals in the search for the point that best separates the categories.

The psychometric background that supports the AQ-Test defines it as a psychometric test characterized by being an instrument used in psychology for the measurement of psychological attributes, standardized, composed of selected and organized items, conceived to produce in the individual the recordable reactions for which it was designed. The AQ-Test questionnaire consists of 50 questions, divided into five areas or subscales, and each subscale consists of 10 items, carefully selected from the behavioral characteristics of Wing's "Triad" [1] of autism symptoms (social skills, attention switching, attention to detail, communication, and imagination).

The test result is a number between zero and 50, with the latter value being the end of the scale with the highest burden of autistic characteristics, where the most adverse levels of affect are located. The test has proposed four ranges or cutoff points for autistic characteristics:

AQ1 (0–10) points = Autistic characteristics below the population average.

AQ2 (11–22) points = Average values of autistic characteristics in a normotypical population (the female average is 15 and the male average is 17).

AQ3 (23–31) points = Autistic characteristics above the population average.

AQ4 (32–50) points = Very high level of autistic characteristics. (Asperger's syndrome or high-functioning autism has an average score of 35).

The study population was grouped according to the resulting score when applying the AQ test, into four groups or phenotypes AQ1 = 28.33%, AQ2 = 43.33%, AQ3 = 10.00%, and AQ4 = 18.33%. When comparing the results of the

---

[3] Phenotypic plasticity has been defined as the capacity of the human brain to change its morphological patterns, interactions in its sensory modalities, and patterns of neurotransmitter secretion. These changes arise by virtue of external and internal factors and represent a phenomenon called neuroplasticity [68, 69].

[4] As the author himself points out, the term quotient is not used here in the arithmetic sense of dividing one quantity by another, but rather derives from the Latin word "quotients," meaning "how much."

Correlation (r) ACC vs. PCC	NAA + NAAG r; p	NAA r; p	Glu + Gln r; p	Glu r; p	GPC + PCh r; p	Cr + PCr r; p	mI r; p	GSH r; p
AQ1	0.14; n.s	0.18; n.s	−0.13; n.s	−0.11; n.s	0.11; n.s	−0.40; n.s	0.26; n.s	0.10; n.s
AQ2	0.14; n.s	0.16; n.s	−0.08; n.s	−0.02; n.s	0.07; n.s	−0.45; 0.02	0.23; n.s	0.05; n.s
AQ3	0.82; 0.05	0.70; n.s	−0.65; n.s	0.03; n.s	−0.54; n.s	0.51; n.s	0.57; n.s	0.41; n.s
AQ4	0.06; n.s	0.25; n.s	0.03; n.s	0.13; n.s	0.58; n.s	0.04; n.s	0.31; n.s	−0.04; n.s

Table 5.
Metabolite Pearson correlation by AQ score (AQ1, AQ2, AQ3, and AQ4) in ACC vs. PCC regions for N-acetyl aspartate (NAA), N-acetyl aspartate +N-acetyl aspartil glutamate (NAA + NAAG), Glutamate+Glutamine (Glx), Glutamate (Glu), creatine (Cr + PCr), Choline (GPC + PCh), myoinositol (mI), and Glutathione (GSH). Regions = Anterior cingulate cortex (ACC) and posterior cingulate cortex (PCC). $p < 0.05$ significantly differences. $P < 0.05$ significant correlation. No significant correlation = n.s. Pearson correlation = r.

spectroscopies reported in point 4.1 where the subjects were grouped according to their neuropsychiatric reports as ASD and TD, with the results obtained by grouping them according to the AQ-test score, significant differences were observed in the absolute concentrations of the neurometabolites between the four groups AQ1, AQ2, AQ3, and AQ4 suggesting a pattern of neuroplasticity [70, 71] characteristic of these disorders.

The results show the absolute concentrations (mM) of the metabolites detected in ACC of the AQ2 group (11–22, mean values of the neurotypical population), AQ3 (23–31, above average) and AQ4 (32–50, very high rate of autistic characteristics) groups compared to the AQ1 group (0–10, below the mean), as well as in the PCC, revealing a specific neurometabolic secretion pattern present in each of the AQ1, AQ2, AQ3, and AQ4 groups (see **Table 4**). Likewise, these results suggest a better fit, when referring to the entire broad spectrum of autism, which could explain the diversity of developmental characteristics in people with this disorder and its complication when making a diagnosis.

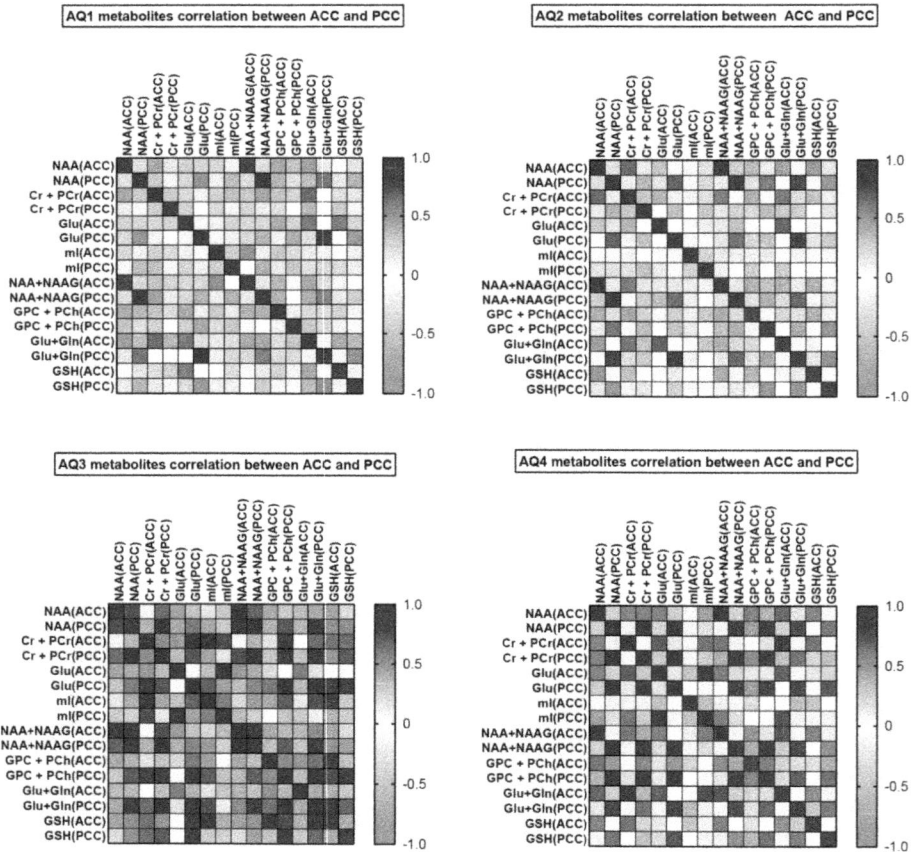

Figure 5.
Heatmap correlation phenotypic neuroplasticity in ACC and PCC within the autism spectrum according to the AQ. Absolute concentrations (mM) of the detected metabolites, for N-acetyl aspartate (NAA), N-acetyl aspartate +N-acetyl aspartil glutamate (NAA + NAAG), Glutamate+Glutamine (Glx), Glutamate (Glu), creatine (Cr + PCr), Choline (GPC + PCh), myoinositol (mI) and Glutathione (GSH). Regions = Anterior cingulate cortex (ACC), and posterior cingulate cortex (PCC). P < 0.05 significantly differences.

By correlating the absolute concentrations of metabolites in ACC vs. PCC, a significant negative correlation was observed for (Cr + PCr) in the AQ2 ($r = -0.45$; $p = 0.02$) group. However, AQ3 highlights that the concentration of (NAA + NAAG) has shown a significant positive correlation in ACC vs. PCC ($r = 0.82$; $p = 0.05$), showing a new promising marker for the detection of ASD by spectroscopy (see **Table 5**).

With these results, we graphed the neurometabolic pattern corresponding to metabolite correlation in ACC vs. PCC in adults, characterized according to the AQ-Test, which made us suggest the great influence due to metabolic dysfunction within the broad spectrum of autism and its relationship with the diversity of characteristics observed in subjects who develop this disorder (see **Figure 5**).

It is necessary to remember that NAAG is distributed together with different neurotransmitters, including glutamate (Glu), known to be a neurotransmitter with excitatory and inhibitory activity in the central nervous system, and gamma-aminobutyric acid (GABA), known as the most common inhibitory neurotransmitter in the CNS. Furthermore, imbalanced (NAAG) metabolism has been described before in some neurological conditions such as schizophrenia, amyotrophic lateral sclerosis, traumatic brain injury, and now in ASD [68, 69, 72].

It caught our attention that although the AQ2 group is made up of subjects with a range of AQ-Test scores (11–22) that are the average values of neurotypical population according to the work carried out by Barón-Cohén; however, neurometabolically, variabilities were observed in the concentrations of the metabolites that make us suggest a dysfunction between ACC and PCC in these subjects without them having the diagnosis of ASD. This does not mean that they do not have some of the characteristics that would identify them as ASD, but that at the neurometabolic and functional level they would show some cognitive deficiencies compared to control group AQ1.

Furthermore, dysregulation of metabolism (NAA + NAAG) between ACC and PCC may also affect the visual system in ASD, highlighting that the posterior cingulate cortex is one of the cortical structures involved in visual attention [73].

5. Conclusions

The pattern we observed in the Heatmap reveals an abnormal neuroplasticity process within the broad spectrum of autism caused by a malfunction of the signaling between neurons in ACC vs. PCC, when observing the neurometabolic correlation pattern of each neurometabolite (see **Figure 5**). The onset of a metabolic change in AQ2, AQ3, and AQ4 is finely detectable when compared to the group of subjects below the average of autistic characteristics AQ1. This study is carried out in adult subjects, which leads us to think that the aberrant neuroplasticity observed in children [74] with ASD continues with them into adulthood.

Thus, the need to perform a correct neurometabolic diagnosis and subsequent treatment in children with ASD is increasingly imminent, recognizing that the same neurometabolites that are treated in the different diseases mentioned throughout this chapter are also altered in ASD. Namely the amino acids and neuropeptides with their derivatives involved in this neuroplasticity process, including glutamate (Glu), glycine (Gly), γ-aminobutyric acid (GABA), and N-Acetyl aspartyl glutamate (NAAG) among others (see **Table 4**). 1H-MRS studies in ASD can be expected to continue to gain importance. Overall, the knowledge gained about neuroplasticity can be applied in clinical and educational areas where all developmental disabilities from childhood onwards underline. Increased knowledge on the neurochemical alteration of autism could lead to a completely new approach to the pharmacological management of autism and to the identification of biomarkers with greater specificity and sensitivity.

Author contributions

C.J.-E. contributed to writing—review and editing, writing—original draft, and research. F.M.S. contributed to writing—review and editing and software. J.L.G-M. contributed to review and editing, supervision, methodology, and conceptualization.

Funding

The author(s) declare that they have received financial support for the research, authorship, and/or publication of this article. This work was funded by the Ministry of Science, Innovation and Universities (grant no. PID2021-126172NB-I00).

Institutional review board statement

This study was carried out in accordance with the recommendations and guidelines set by the European Union (86/609/EU) and in accordance with the current legal regulation of Spain Law 32/2007, RD 53/2013, Law 9/2003, and RD 178/2004. The protocol research was approved by the Ethics Committee of La Laguna University and Ministry of Science and Innovation (Spain), with approval number CEIBA2013–0056. This study was carried out in accordance with the recommendations of the ethical principles for medical research involving human, and law 14/2007 for biomedical research, with written information from all subjects. The study was conducted according to the guidelines of the Declaration of Helsinki and approved by the Institutional Board.

Informed consent statement

Informed consent was obtained from all subjects involved in the study.

Conflict of interest

The authors declare no conflict of interest.

Author details

Carmen Jiménez-Espinoza*, Francisco Marcano Serrano and José González-Mora
Neurochemistry and Neuroimages Laboratory, Department of Basic Medical
Sciences, Faculty of Health Sciences, Physiology Section, University of La Laguna,
Tenerife, Spain

*Address all correspondence to: carmen.jimenez.87@ull.edu.es

IntechOpen

References

[1] Wing L, Gould J. Severe impairments of social interaction and associated abnormalities in children: Epidemiology and classification. Journal of Autism and Developmental Disorders. 1979;**9**(1):11-29

[2] American Psychiatric Association et al. Diagnostic and statistical manual of mental disorders. Text Revision. 2000

[3] Clarke PHD, Dudley D, Sokoloff L. Circulation and Energy Metabolism in the Brain/Donald D. Clarke and Louis Sokoloff; 1999

[4] Siesjö BK. Brain energy metabolism and catecholaminergic activity in hypoxia, hypercapnia and ischemia. Journal of Neural Transmission. Supplementum. 1978;**14**:17-22

[5] Teleanu RI et al. Neurotransmitters— Key factors in neurological and neurodegenerative disorders of the central nervous system. International Journal of Molecular Sciences. 2022;**23**(11):5954

[6] Hetherington HP, Gadian DG, Ng TC. Magnetic resonance spectroscopy in epilepsy: Technical issues. Epilepsia (Series 4). 2002;**43**

[7] Beckwith-Hall BM et al. Nuclear magnetic resonance spectroscopic and principal components analysis investigations into biochemical effects of three model hepatotoxins. Chemical Research in Toxicology. 1998;**11**(4):260-272

[8] Baslow MH. Functions of N-acetyl-L-aspartate and N-acetyl-L-aspartylglutamate in the vertebrate brain: Role in glial cell-specific signaling. Journal of Neurochemistry. 2000;**75**(2):453-459

[9] Pioro EP et al. Detection of cortical neuron loss in motor neuron disease by proton magnetic resonance spectroscopic imaging in vivo. Neurology. 1994;**44**(10):1933-1933

[10] Courvoisie H et al. Neurometabolic functioning and neuropsychological correlates in children with ADHD-H: Preliminary findings. The Journal of Neuropsychiatry and Clinical Neurosciences. 2004;**16**(1):63-69

[11] Kreis R, Ernst T, Ross BD. Cuantificación absoluta de agua y metabolitos en el cerebro humano. II. Concentraciones de metabolitos. Journal of Magnetic Resonance, Serie B. 1993;**102**(1):9-19

[12] Castillo M, Smith JK, Kwock L. Correlation of myo-inositol levels and grading of cerebral astrocytomas. American Journal of Neuroradiology. 2000;**21**(9):1645-1649

[13] Govindaraju V, Young K, Maudsley AA. Proton NMR chemical shifts and coupling constants for brain metabolites. NMR in Biomedicine: An International Journal Devoted to the Development and Application of Magnetic Resonance In Vivo. 2000;**13**(3):129-153

[14] Frye RE et al. Redox metabolism abnormalities in autistic children associated with mitochondrial disease. Translational Psychiatry. 2013;**3**(6):e273-e273

[15] Janáky R et al. Glutathione in the nervous system: Roles in neural function and health and implications for neurological disease. In: Handbook of Neurochemistry and Molecular Neurobiology: Amino Acids and Peptides

in the Nervous System. Springer US; 2007. pp. 347-399

[16] Rossignol DA, Frye RE. Mitochondrial dysfunction in autism spectrum disorders: A systematic review and meta-analysis. Molecular Psychiatry. 2012;**17**(3):290-314

[17] Pastore A et al. Analysis of glutathione: Implication in redox and detoxification. Clinica Chimica Acta. 2003;**333**(1):19-39

[18] Darnell J, Lodish H, Baltimore D. Molecular Biology of the Cell. 1990

[19] Garcia-Lopez M. Axolemmal transporters for neurotransmitter uptake. Revista de Neurologia. 1999;**29**(11):1056-1063

[20] Nadim F, Bucher D. Neuromodulation of neurons and synapses. Current Opinion in Neurobiology. 2014;**29**:48-56

[21] Nan F et al. Dual function glutamate-related ligands: Discovery of a novel, potent inhibitor of glutamate carboxypeptidase II possessing mGluR3 agonist activity. Journal of Medicinal Chemistry. 2000;**43**(5):772-774

[22] Yamasue H, Domes G. Oxytocin and autism spectrum disorders. Behavioral Pharmacology of Neuropeptides: Oxytocin. 2018:449-465

[23] Kim D, Lee J-S. Neurotransmitter-induced excitatory and inhibitory functions in artificial synapses. Advanced Functional Materials. 2022;**32**(21):2200497

[24] Chao H-T et al. Dysfunction in GABA signalling mediates autism-like stereotypies and Rett syndrome phenotypes. Nature. 2010;**468**(7321):263-269

[25] Cornell-Bell AH et al. Glutamate induces calcium waves in cultured astrocytes: Long-range glial signaling. Science. 1990;**247**(4941):470-473

[26] Kaufman AS, Raiford SE, Coalson DL. Intelligent Testing with the WISC-V. John Wiley & Sons; 2015

[27] Kaufman NL, Kaufman AS. Comparison of normal and minimally brain dysfunctioned children on the McCarthy scales of children's abilities. Journal of Clinical Psychology. 1974;**30**(1)

[28] Bracken BA et al. Universal nonverbal intelligence test (UNIT). Alternative Assessments With Gifted and Talented Students. 2021

[29] Ling W et al. Construction of CPM scale for leadership behavior assessment. Acta Psychologica Sinica. 1987;**19**(02):89

[30] Baron-Cohen S et al. Early identification of autism by the checklist for autism in toddlers (CHAT). Journal of the Royal Society of Medicine. 2000;**93**(10):521-525

[31] Mazefsky CA, Oswald DP. The discriminative ability and diagnostic utility of the ADOS-G, ADI-R, and GARS for children in a clinical setting. Autism. 2006;**10**(6):533-549

[32] Rutter M, Le Couteur A, Lord C. ADI-R. Autism Diagnostic Interview Revised. Manual. Los Angeles: Western Psychological Services; 2003

[33] Baron-Cohen S. La gran diferencia: Cómo son realmente los cerebros de hombres y mujeres. Editorial AMAT. 2005

[34] Graaf D, Robin A. In Vivo NMR Spectroscopy: Principles and Techniques. John Wiley & Sons; 2019

[35] Provencher SW. Estimation of metabolite concentrations from localized in vivo proton NMR spectra. Magnetic Resonance in Medicine. 1993;**30**(6):672-679

[36] Cramér H. Mathematical Methods of Statistics. Princeton University Press; 1999

[37] Provencher SW. Automatic quantitation of localized in vivo 1H spectra with LCModel. NMR in Biomedicine: An International Journal Devoted to the Development and Application of Magnetic Resonance In Vivo. 2001;**14**(4):260-264

[38] Ross RG et al. Perinatal choline effects on neonatal pathophysiology related to later schizophrenia risk. American Journal of Psychiatry. 2013;**170**(3):290-298

[39] Bejjani A et al. Elevated Glutamatergic Compounds in Pregenual Anterior Cingulate in Pediatric Autism Spectrum Disorder Demonstrated by 1H MRS and 1H MRSI. 2012

[40] Santello M, Bezzi P, Volterra A. TNFα controls glutamatergic gliotransmission in the hippocampal dentate gyrus. Neuron. 2011;**69**(5):988-1001

[41] Petrelli F, Pucci L, Bezzi P. Astrocytes and microglia and their potential link with autism spectrum disorders. Frontiers in Cellular Neuroscience. 2016;**10**:21

[42] Habbas S et al. Neuroinflammatory TNFα impairs memory via astrocyte signaling. Cell. 2015;**163**(7):1730-1741

[43] Murphy KM et al. Development of human visual cortex: A balance between excitatory and inhibitory plasticity mechanisms. Developmental Psychobiology: The Journal of the International Society for Developmental Psychobiology. 2005;**46**(3):209-221

[44] Jimenez-Espinoza C, Serrano FM, González-Mora J. Glutamate Dysregulation in Cingulated Cortices Is Associated with Autism Spectrum Disorder Traits. 2024

[45] Hynd MR, Scott HL, Dodd PR. Glutamate-mediated excitotoxicity and neurodegeneration in Alzheimer's disease. Neurochemistry International. 2004;**45**(5):583-595

[46] Estrada-Sánchez AM et al. Glutamate toxicity in the striatum of the R6/2 Huntington's disease transgenic mice is age-dependent and correlates with decreased levels of glutamate transporters. Neurobiology of Disease. 2009;**34**(1):78-86

[47] Caudle WM, Zhang J. Glutamate, excitotoxicity, and programmed cell death in Parkinson disease. Experimental Neurology. 2009;**220**(2):230-233

[48] Young KW et al. Different pathways lead to mitochondrial fragmentation during apoptotic and excitotoxic cell death in primary neurons. Journal of Biochemical and Molecular Toxicology. 2010;**24**(5):335-341

[49] Murphy DGM et al. Asperger syndrome: A proton magnetic resonance spectroscopy study of brain. Archives of General Psychiatry. 2002;**59**(10):885-891

[50] Cacciaguerra L et al. Neuroimaging features in inflammatory myelopathies: A review. Frontiers in Neurology. 2022;**13**:993645

[51] Bertrand D, Wallace TL. A review of the cholinergic system and therapeutic approaches to treat brain disorders. Behavioral Pharmacology of the Cholinergic System. 2020:1-28

[52] Mesulam M-M. The cholinergic innervation of the human cerebral cortex. Progress in Brain Research. 2004;**145**:67-78

[53] Chan AS et al. Disordered connectivity associated with memory deficits in children with autism spectrum disorders. Research in Autism Spectrum Disorders. 2011;**5**(1):237-245

[54] Robillard R et al. Lower in vivo myo-inositol in the anterior cingulate cortex correlates with delayed melatonin rhythms in young persons with depression. Frontiers in Neuroscience. 2017;**11**:336

[55] Pettem KL et al. Interaction between autism-linked MDGAs and neuroligins suppresses inhibitory synapse development. Journal of Cell Biology. 2013;**200**(3):321-336

[56] Sani A, Fathiah N, et al. Gene expression profile in different age groups and its association with cognitive function in healthy Malay adults in Malaysia. Cells. 2021;**10**(7):1611

[57] Tay TL et al. Microglia across the lifespan: From origin to function in brain development, plasticity and cognition. The Journal of Physiology. 2017;**595**(6):1929-1945

[58] Southwick JS et al. Memory functioning in children and adolescents with autism. Neuropsychology. 2011;**25**(6):702

[59] Leech R, Braga R, Sharp DJ. Echoes of the brain within the posterior cingulate cortex. Journal of Neuroscience. 2012;**32**(1):215-222

[60] Mazengenya P, Bhikha R. An analysis of the structural and functional anatomy of the spine as viewed by Avicenna (980—1037 AD). Research Journal of Medical Sciences. 2016;**1**:676-682

[61] James SJ et al. Metabolic biomarkers of increased oxidative stress and impaired methylation capacity in children with autism. The American Journal of Clinical Nutrition. 2004;**80**(6):1611-1617

[62] Melnyk S et al. Metabolic imbalance associated with methylation dysregulation and oxidative damage in children with autism. Journal of Autism and Developmental Disorders. 2012;**42**:367-377

[63] Rose S et al. Evidence of oxidative damage and inflammation associated with low glutathione redox status in the autism brain. Translational Psychiatry. 2012;**2**(7):e134-e134

[64] Jiménez-Espinoza C, Serrano FM, González-Mora J. Imbalance Glutathione Biosynthesis in ASD: A Kinetic Pattern "in Vivo". 2018

[65] Rossignol DA, FRYE, Richard E. Evidence linking oxidative stress, mitochondrial dysfunction, and inflammation in the brain of individuals with autism. Frontiers in Physiology. 2014;**5**:150

[66] Rossignol DA, Genuis SJ, Frye RE. Environmental toxicants and autism spectrum disorders: A systematic review. Translational Psychiatry. 2014;**4**(2):e360-e360

[67] Baron-Cohen S et al. The autism-spectrum quotient (AQ): Evidence from Asperger syndrome/high-functioning autism, males and females, scientists and mathematicians. Journal of Autism and Developmental Disorders. 2001;**31**:5-17

[68] Neale JH et al. The neurotransmitter N-acetylaspartylglutamate in models of pain, ALS, diabetic neuropathy, CNS injury and schizophrenia.

Trends in Pharmacological Sciences.
2005;**26**(9):477-484

[69] Tsai S-J. Central N-acetyl
aspartylglutamate deficit: A possible
pathogenesis of schizophrenia.
Medical Science Monitor.
2005;**11**(9):HY39-HY45

[70] Buonomano DV, Merzenich MM.
Cortical plasticity: From synapses to
maps. Annual Review of Neuroscience.
1998;**21**(1):149-186

[71] Debanne D et al. Brain plasticity and
ion channels. Journal of Physiology-Paris.
2003;**97**(4-6):403-414

[72] Jiménez-Espinoza C, Marcano
Serrano F, González-Mora JL.
N-acetylaspartyl-glutamate metabolism
in the cingulated cortices as a biomarker
of the etiology in ASD: A 1H-MRS model.
Molecules. 2021;**26**(3):675

[73] Rademacher J et al. Human cerebral
cortex: Localization, parcellation,
and morphometry with magnetic
resonance imaging. Journal of Cognitive
Neuroscience. 1992;**4**(4):352-374

[74] Chen Z et al. Neuroplasticity of
children in autism spectrum disorder.
Frontiers in Psychiatry. 2024;**15**:1362288